AF774701

BLOCKCHAIN INTELLIGENT SYSTEMS

Protocols, Application and Approaches for Future Generation Computing

Editors

E. Golden Julie
Assistant Professor [Sr. Grade], Department of CSE
Anna University, Regional Campus, Tirunelveli, India

Y. Harold Robinson
Professor, Department of CSE
Francis Xavier Engineering College
Vannarpettai, Tirunelveli-627003

J. Jesu Vedha Nayahi
Assistant Professor (Sr. Grade), Department of CSE
Anna University, Regional Campus, Tirunelveli-627007

Thavavel Vaiyapuri
Assistant Professor, College of Computer Engineering and Sciences
Prince Sattam bin Abdulaziz University, Alkharj, Saudi Arabia

CRC Press
Taylor & Francis Group
Boca Raton London New York

CRC Press is an imprint of the
Taylor & Francis Group, an **informa** business
A SCIENCE PUBLISHERS BOOK

First edition published 2024
by CRC Press
2385 NW Executive Center Drive, Suite 320, Boca Raton FL 33431

and by CRC Press
4 Park Square, Milton Park, Abingdon, Oxon, OX14 4RN

© 2024 E. Golden Julie, Y. Harold Robinson, J. Jesu Vedha Nayahi and Thavavel Vaiyapuri

CRC Press is an imprint of Taylor & Francis Group, LLC

Reasonable efforts have been made to publish reliable data and information, but the author and publisher cannot assume responsibility for the validity of all materials or the consequences of their use. The authors and publishers have attempted to trace the copyright holders of all material reproduced in this publication and apologize to copyright holders if permission to publish in this form has not been obtained. If any copyright material has not been acknowledged please write and let us know so we may rectify in any future reprint.

Except as permitted under U.S. Copyright Law, no part of this book may be reprinted, reproduced, transmitted, or utilized in any form by any electronic, mechanical, or other means, now known or hereafter invented, including photocopying, microfilming, and recording, or in any information storage or retrieval system, without written permission from the publishers.

For permission to photocopy or use material electronically from this work, access www.copyright.com or contact the Copyright Clearance Center, Inc. (CCC), 222 Rosewood Drive, Danvers, MA 01923, 978-750-8400. For works that are not available on CCC please contact mpkbookspermissions@tandf.co.uk

Trademark notice: Product or corporate names may be trademarks or registered trademarks and are used only for identification and explanation without intent to infringe.

Library of Congress Cataloging-in-Publication Data (applied for)

ISBN: 978-1-032-32220-9 (hbk)
ISBN: 978-1-032-32221-6 (pbk)
ISBN: 978-1-003-31344-1 (ebk)

DOI: 10.1201/9781003313441

Typeset in Palatino Linotype
by Radiant Productions

Preface

In recent years block chain is one of the advanced topics for both academic and industry. This block chain will give vast opportunities for the development of future generation computing. combine artificial intelligence with block chain technology will improve the future generation computing in all aspects. The basic concept behind Intelligence is that a system that functions as an electronic butler that senses user features and their environment can develop by enriching the environment with technology, such as sensors and devices that are networked. This book aim to describe details about block chain and Application area of block chain. Block chain and security systems provide a new way for innovating a wide range of application scenarios due to their beneficial characteristics of tamper-proof, transparency, trustworthy, and privacy protection. It can also ensure security and reduce property losses by enabling anonymous and trustful transactions in trust less environments. The recent boom of block chain systems provides new opportunities and challenges, such as the lack of unified system performance evaluation indicators, heterogeneous block chain systems are difficult to interact with, K-V data storage structure leads to single data query functions and fraudulent smart contracts, and so on. Intelligent block chain is expected to play as a game-changing role in future communications and networking, since it can be exploited to provide innovative solutions to lots of emerging research issues in the next-generation communications and networking.

Chapter 1 describe about social theory in block chain with societal impact. This chapter begin with introduction and evolution of block chain. Then followed by ethical framework for block chain. In Opportunities and Challenges ahead have discussed coin mining, social impact of block chain technology, trust and code and coder, etc.

Chapter 2 Discussed about compactable works in block chain Technology. It starts with introduction, IoT with block chain, Security and Privacy issues and Finally learning scheme in block chain.

Chapter 3 covers the medical application of an Integration of IoT and Block chain. This chapter begin with introduction about IoT in medical field. Followed by detail description and types of block chain in medical field. Then combination of IoT and Block chain in health care application. Finally conclude with limitation and future scope.

Chapter 4 covers the medical application using block chain introduction part long with need of block chain in healthcare area. Followed by support of healthcare in global along with future of block chain in health care application. Focus on challenges and integration of IoT along with block chain. Discussed about Healthcare Data sharing Through the Gem Health Network. Finally discussed about future scope of block chain along with security issues.

Chapter 5 covers the topic as deal with block chain and IoT for health care Applications. Followed by comparative analysis of block chain and IoT in healthcare applications. Discussed the scenario for healthcare application challenges and case study was finally discussed.

Chapter 6 deals with secure E-Health system using block chain technologies in IoT Environment. It begins with introduction, need for health care in Block chain, Working Technology, followed by cloud computing in Health care. Data Extraction in IoT based cloud computing. Finally, architecture concepts of block chain with conclusion.

Chapter 7 Discussed about Machine learning model for predicting viral diseases using Linear regression. It begins with introduction, Related work, Regression for predicting viral diseases, Data set used and Discussed about experimental result with conclusion.

Chapter 8 Focus about Ethereum based electronic health record storage. It discussed detail about introduction about block chain and structure of block chain. The main discussion of the chapter is taxonomy of Electronic health records. It shows the architecture view of the electronic health record storage. It finally discussed about algorithm with result discussion.

Chapter 9 This chapter provides detail privacy prevention system for the Internet of Medical Things. It starts discussing about Introduction, key generation for privacy preserving, Pseudocode for hybrid hashing for BIoMT, Smart contract. Finally discussed about result of the implementation with conclusion.

Chapter 10 describe the block chain based E-voting system. Begin with introduction about block chain architecture. Followed by Election role and process along with Evaluation and implementation. Finally discussed bout security requirements for voting system along with benefits and disadvantages.

Chapter 11 covers the prediction of student's performance in secure and effective manner using block chain and data mining. The chapter begin with discussing about introduction followed by related work along with methodology. Such as data collection, Data storage, etc. Finally discussed along with result.

Acknowledgment

"Success is the sum of small efforts repeated day in and day out."
—Robert Collier statement

The successful completion of this edited book is possible with the help from various resources. We express our sincere thanks to all of them.

We feel deeply indebted to Almighty God for giving this opportunity. I extend my deep sense of gratitude to our Son Master H. Jubin, and family members Mr. S. Eanoch, Mrs. E. Lizzy Pushpa Bai, Mr. K. Yesudhas and Mrs. Y. Heln Pushpa Rathi for their moral support and encouragement at all stages for the successful completion of this book.

We extend our deep sense of gratitude to scholars and friends for writing their chapter in time. We sincerely thank our friends and family member for their necessary prayer support.

Finally, We would like to take this opportunity to specially thank CRC Press for their help, encouragement and moral support.

E. Golden Julie and Y. Harold Robinson

Contents

Chapter 1

Blockchain Social Theory and Societal Impact

Danilo Pelusi,[1] *Yuvalatha S,*[2,*] *Bala Preethika J,*[2] *Jeya Brundha K*[2] and *Vignesh Srinivasan S*[2]

We face a daily reality such that very interesting outlook changes are testing the fundamental organizations that trust is based on. The absence of trust related with information administration, which is frequently knowledgeable about the type of information breaks or essentially an adaptation of our information without our consent as well as impetus to take part in this arising decentralization of designs, is basically difficult centralization as states, monetary establishments, ventures, and associations. We perceive this trust gap posing a threat to the very institutions on which we have relied, such as financial institutions, private businesses, and government agencies. As people keep on advancing into more decentralized and self-administering (or semi-independent) associations, a new "common agreement" is fundamental. Although there has been significant discussion regarding blockchain applications and prospective results in the FinTech industry, little has been done to examine how client-driven blockchain advancements can enable a variety of purposes outside of banking. This article plans to

[1] Faculty of communication Sciences, University of Teramo Italy.

[2] Bannari Amman Institute of Technology, Sathyamangalam.

Emails: dpeslusi@unite.it; balapreethika.cb20@bitsathy.ac.in; jeyabrundha.cb20@bitsathy.ac.in; vigneshsrinivasan.cb20@bitsathy.ac.in

* Corresponding author: yuvalathas@bitsathy.ac.in

add to that assortment of information by inspecting blockchain innovation's likely applications, as well as its restrictions, in regions where social effect crosses, like basic liberties. This likewise examines whether blockchain innovation and its center functional standards - like decentralization, straightforwardness, equity, and responsibility - can assist with restricting unnecessary internet based observation, oversight, and common liberties infringement that are worked with by the developing dependence on a couple of elements to control admittance to data on the web. With regards to the expected effect of blockchain innovation on society, what is conceivable and what ought to be kept away from. With the proviso that an administration instrument should be indicated, we will see further improvement in computerized power. The data introduced in this position paper upholds the possibility that blockchain and individual tokenization could make another common agreement.

1. Introduction

Blockchain started as the conveyed record innovation (DLT) that supports Bitcoin and other digital currencies. Blockchain has arisen in an assortment of enterprises as of late, including monetary advances (FinTech), sharing economies, medical services, science, government, and regulation. Notwithstanding the new amazing development of blockchain, genuine utilization is as yet restricted. Subsequently, legislatures and different associations are as yet watching out for block chain's advancement. Concerns about blockchain and its applications continue to exist. It is necessary to address the moral and ethical issues. Without the use of banks or third-party platforms, blockchain technology has enabled the creation of a worldwide functional currency based on code. The Internet of Things (IoT) and Fifth-Generation Technology (5G) are two huge fields that will drive decentralization somewhat. Supply chains, for instance, are inherently confounded, need straightforwardness, and result in duplication of obligations across partners, diminishing proficiency and inflating costs. Instead of a tokenization economy, we are currently experiencing the social democratization of information. Self-tokenization and information democratization are two new ways to deal with forestalling information spillage and safeguarding individual data. The uses of this internet-based distributed ledger technology now range from simple digital identity verification to automating multilayered payments utilizing complex smart contracts. The guarantee of a future wherein shoppers straightforwardly direct administrations and never again depend on strong go-betweens rouses colossal excitement, yet additionally incredible fear, which reaches out a long ways past the FinTech1 business to regions that might hurt individuals' occupations and freedoms. One arrangement that can possibly further develop decentralization, straightforwardness, correspondence, and responsibility on the web is blockchain innovation. However, one must understand that the technology is still relatively unproven and may contain both known and unknown hazards, which we will discuss in this essay. Blockchain innovation, as other secrecy based advancements, can be utilized for both great and malicious purposes, for example, facilitating unlawful substance and exercises. We are currently experiencing the social democratization of information rather than a tokenization economy. This article intends to add to that group of information by analyzing blockchain innovation's expected applications, as well as its impediments, in regions where social effect converges, like basic liberties. This article explores whether blockchain innovation and its center functional standards - like decentralization, straightforwardness, correspondence, and responsibility - can assist with restricting unnecessary internet based reconnaissance, oversight, and basic liberties infringement that are worked with by the developing dependence on a couple of substances to control admittance to data on the web. Therefore, this paper means to provoke insightful curiosity in what the future held can make concern when it comes the likely effect of blockchain innovation on society.

2. Research Background

Blockchain began under 10 years prior as the DLT that upholds bitcoin [2]. By establishing a trustless climate, blockchain permits steady, powerful, calculation guaranteed information capacity and exchange approval, as well as independent handling. Like the instance of PC morals [1], comprehension of the fundamental specialized ideas is required for an unmistakable comprehension of the morals of blockchain. The accompanying area presents the focal ideas basic the blockchain innovation.

2.1 *Evolution of Blockchain*

Bitcoin, a form of digital currency released in 2009 as an open-source framework by a person or group of persons going by the name Satoshi Nakamoto, was the first to use blockchain technology as the underlying structure and mechanism. Cryptocurrencies have grown in popularity due to improvements in computer power, accessibility, and cost, but their popularity must be understood in the context of a significant public backlash against centralized systems and organizations in the wake of the 2008 financial crisis. This emergency, which a few business analysts contrast with the Great Depression, broke individuals' confidence in customary monetary mediators. Bitcoin gave a mechanical arrangement: interestingly, individuals would have the option to manage huge scope financial exchanges without trusting or depend on outsiders. Blockchain interconnected chain of squares where each square is a record of a cryptographically marked exchange. This is why the term "blockchain" refers to a conveyed computerized record or bookkeeping book.

3. Main Concepts

3.1 Encryption

Advanced change, which incorporates "cutting edge", for example, AI and square chains, has opened up a huge number of new choices for organizations to develop by expanding purchaser esteem and conveying expanded efficiencies to a different gathering of clients. As a result of this transition, both individuals and organizations have benefited greatly, but they have likewise become defenseless against security gambles and digital culpability. The worldwide monetary market breakdown delivered remarkable financial headwinds, yet it likewise agreed with the absolute most prominent information breaks in history. As a matter of fact, the examination observed that the breaks might have been totally stayed away from assuming fitting security techniques had been set up at the hour of the occurrence. Security consistence is an asset concentrated challenge for all organizations, paying little mind to measure. In spite of associations' earnest attempts and enormous interests in information security, countless records have been penetrated and keep on being scrambled. For instance, the Verizon Business RISK group announced in 2008 that 3/4 of all breaks happened in retail, monetary administrations, and food and drink. Servers and programs were used to breach the majority of the records. In 66% of the issues, the break compromised information that the organization was ignorant was put away on its frameworks. Since 2008, various industries have been plagued by data leaks. In 2018, there were an unprecedented amount of data breaches of varied severity. In India, a hacking bunch put malevolent code into British Airways' inadequately gotten website pages, uncovering the information of 380,000 clients, and an ID data set claimed by the Indian government was hacked, uncovering the distinguishing proof subtleties and private data of 1.1 billion individuals. Encryption and tokenization are two advances that are utilized to safeguard information. Tokenization and encryption are every now and again spoken in a similar sentence, and the terms are utilized conversely. While both are utilized to jumble information, there are massive contrasts between the two. The two advances can be utilized to safeguard information in various circumstances, and sometimes, for example, start to finish installment assistance. A similar key is utilized to encode and decode data in symmetric key encryption. In erroneous key encryption, encryption and decryption are performed using two distinct keys. Therefore, communications sent to a single person can be decoded using the private key, but messages sent to another segment that are encoded using a different key pair cannot. To protect data communicated over the Internet, Secure Attachments Layer (SSL) encryption is currently widely utilized. A great many clients scramble information on their PCs utilizing the underlying encryption abilities of OS frameworks or outsider encryption applications to defend delicate information from being coincidentally lost assuming their machine is taken. Scrambling information is a tedious and asset concentrated process.

3.2 *Tokenization*

Then again, tokenization is the most common way of changing a piece of information into an arbitrary series of characters known as a token. Tokenization substitutes delicate information with non-touchy information to safeguard touchy information. The token simply fills in as a pointer to the first information; computing the values can't be utilized. Inside the dealer's stockpiling climate, a charge card number, for instance, is supplanted with a symbolic worth that can't be followed back to the first information piece. Accordingly, the symbolic worth can be utilized as a substitute for genuine information in an assortment of uses, some of which are talked about in this paper. Tokens enjoy the benefit of having no numerical relationship to the hidden information they address. Since the genuine information values can't be recuperated through opposite, a break makes the data incredibly important. Tokens are progressively being used to safeguard different kinds of touchy information. Individual distinguishing data, for example, medical care data, email locations, and record numbers, are instances of this kind of information. Tokenization significantly limits risk from a security point of view since touchy information can't be penetrated in the event that it is absent in any case. Other use situations where tokens produce esteem incorporate the tokenization of customary monetary resources where liquidity makes obstacles to section, and this problem can be solved by tokenizing these resources, for as by converting freedoms into a digital token supported by the real resource utilizing blockchain. We are supposed to observe a flood in the quantity of tokenized genuine resources, for example, land and collectibles like workmanship, where financial backers can claim bits of land and collectibles. Personalization and customization will turn out to be more pervasive later on. The token reflects scattered responsibility for fundamental resources worth, not simply the resource, in the two cases, democratizing the proprietorship process. The potential for a new "tokenized" economy to establish a more productive and comprehensive climate where unmistakable and elusive resources can be exchanged with more noteworthy liquidity, openness, straightforwardness, and quicker and savvier exchanges is gigantic. Moreover, tokenization permits you to effectively and actually compartmentalize individual information and oversee it across a few clients. Therefore, just a substance with the significant token can get to that information. Since individual information is overseen by the client and imparted to authorization, the data is secure and very much versatile, rather than information that is at present utilized.

4. Increased Centralization

The expanded interest in and utilization of dispersed record advances, for example, the blockchain should be grasped interestingly, with the high centralization of client connections and information at the web's application layer, like on web indexes, interpersonal organizations and content stages [7]. Helped by the ascent of cell phones and less expensive, simpler media transmission access, the web has seen huge development starting around 2000. Simultaneously, web clients have step by step went to a somewhat little arrangement of conglomerating stages to recover data and speak with each other [11]. After some time, the web turned out to be more concentrated, which is somewhat unreasonable since decentralization is the foundation upon which the internet was built. This development may be seen in the fact that the wealth accumulated over the last ten or so years mostly benefited a small number of international partnerships, whose system was to link online users by acting as a client interface. The largest taxi company in the world, Uber, says it has no vehicles. The most well-known media company in the world, Facebook, does not make people happy. The largest retailer, Alibaba, has no inventory. In addition, Airbnb, the largest provider of conveniences worldwide, makes no rights to any land [3]. Billions of individuals from all areas of the planet have become subject to the computerized items and administrations of few worldwide companies that serve, whether they need to or not, as stifle focuses for people's very own information, discourse from there, the sky is the limit [2]. While web clients remain exceptionally disseminated on a worldwide organization of organizations, the information these clients produce are a lot of moved in the possession of a couple of organizations. Customers benefit from this interconnection since it reduces the time and effort required to access their information or use other web-based services. It also enables several private businesses to thrive on such platforms and provides some web users, like video bloggers, with the beneficial opportunity to make income by altering content through promotions and other approaches [11]. However, having such a high degree of centralization is the exact antithesis of the web's original commitment to being an open-source, interoperable organisation that can endure terrible disasters without losing its fundamental actual organisation [8]. The creator of the World Wide Web, Tim Berners-Lee, has publicly demanded that one of the decentralisation movement's original notions be revisited: the web is now decentralised. The problem is the dominance of one big informal community, one web search engine, and one microblogging platform like Twitter. We have a societal problem, not an innovative problem [8]. A few recent cyberattacks and targeted setbacks have shown that the concentrated approach is defenceless against a variety of misuse or error that might harm people, including the right to security and opportunity of expression. For instance, the Snowden revelations revealed that the United States government has direct access to information from Google, Facebook, and other US companies (Landau). Another example is the 2012 LinkedIn hack, which resulted in the compromising of more than 100 million user accounts [9]. Information from Freedom House suggests that web restrictions, such as oversight, content separation, and blogger provocation, to name a few, have been increasing since at least 2012, which is in favour of the chance for articulation (Opportunity House 2016). It is crucial to keep in mind that in the middle of the 2000s, several attempts were made to lessen the severity of the centralization of online content services using dispersed (P2P) apps, such as, Napster and conventions, for example, Bit Torrent, which aided battle centralization by utilizing overlay organizations like Free net, an exceptional strategy for restricting observation and control [1]. However, utilizing such devices additionally brought about legitimate difficulties, predominantly in the area of copyright encroachments [10]. A few different innovations zeroed in more on restricting reconnaissance by network access suppliers (ISPs) and states. They did as such by keeping away from the utilization of organizations and on second thought by depending on different clients as intermediaries. One such initiative is the Tor Project, one of the most active open-source initiatives aimed at hindering safe online insight. However, each of them nevertheless had its limitations in terms of man-in-the-center attacks or revealing some of the company's clients [13]. Besides, a response from the weak against strong mediators can't alone make sense of the early reception and current purposes of blockchain or other conveyed advancements. There are a few models where criminal operations were a calculate utilizing such innovations. For example, BitTorrent was found to have been utilized to trade kid sexual entertainment [10], Tor was ascribed with making commercial centers for illegal exploitation and different violations and Bitcoin was the digital money utilized by programmers to coerce assets from casualties as a trade-off for reestablishing admittance to significant information [4]. As such, remembering that is significant innovation isn't innately positive or negative; rather it makes a bunch of possibilities that are informed by its center engineering and standards. Against this setting, let us make a plunge in additional detail with a portion of the commitments and entanglements of blockchain innovation with regards to its effect on society at large.

5. Decentralization

It is basic to characterize language and setting while at the same time examining the shift toward "decentralization." An assemblage of examination focused on the investigation of association figuring has been conveyed inside the setting of associations. Cooperative interchanges are worked with by the basic organization and IT applications. P2P design alludes to an essential organization of hubs wherein clients don't share any of their assets, for example, extra room, handling power, network association, data transfer capacity, or content. Decentralization in this study might infer the use of conveyed and distributed abilities. The monetary emergency of 2008, different breaks of resident information as hacks, and a deficiency of security are for the most part contributing variables to the shift toward decentralization on a social level. The centralization of citizen data, according to the theory, can be abused. This centralization can occur in government and private institutions where the governance concept is the same. Because of these abuses, it's debatable. "Information is the new oil," as the maxim goes, only adds to the worth of resident information and the translation that can be produced using it. A shrewd agreement is a PC program code that computerizes the most common way of finishing an exchange and is equipped for working with, doing, and authorizing the understanding. Using blockchain technology, the performance of an agreement is measured. Theory of the Social Contract may best be described as the process of creating policy consensus among designated authority. Government agencies and citizens, for example. It's implied that they have similar morals and ideals. Common agreements have advanced to personality legislative issues and the idea of numerous characters in the twenty-first century, helped by the utilization of online entertainment and the Internet, especially with regards to IoT and the Internet of Everything and Everyone. The requirement for another social minimized that can work in an advanced climate is being provoked by pride and the battle for affirmation. Given the decentralization pattern or a hybrid thereof, an innovative and meaningful social contract must place the citizen as an active player at the center of the digital universe. In a carefully determined world, a resident may be depicted as a person who specifically and deliberately states their privileges. As a result, a citizen does not always imply a person who is a citizen of a certain country, nor does it imply a global citizen. An esteemed computerized resident should be the wellspring of the common agreement among him and the substances he "trusts," ideally in decentralized modalities. This is a clever case, considering that common agreements in the twenty-first century have advanced to incorporate everything from general fundamental pay to the privilege to guaranteed work under Industry 4.0. Where common agreements consolidate the instrument of assessments that is expected to give "aggregate government assistance" for country state individuals, there is still a reliance on government organizations. Who makes the rules and how they are executed could be determined by a social smart contract. Over the last ten years, the blockchain has evolved to incorporate enterprise and, in some circumstances, government usage. An advanced resident driven social and brilliant agreement, then again, has been the missing connection. There is a typical conviction that the computerized resident has earlier information on both turn of events and execution. Utilization of arising components that advance the development of another social request contract. This supposition could be right; consequently, more research into the subject is necessary. It is suggested that certain types of training and education be provided, as well as by particular organizations.

6. Smart Contracts

All blockchain-based services have an architecture that is focused on immutability and decentralization in data storage. Nonetheless, a local area based framework can't work without the local area. Each blockchain-based assistance has its own local area, which changes relying upon how it is constructed and run. While Bitcoin is dependent on brokers, clients, diggers, and center engineers, each help has own local area of individuals cooperate to make, create, and use it. Brilliant agreements, which may be described as automated computer programs that can be activated to move digital resources in the corresponding blockchain, while specific setting off conditions are met, are one of the most up to date parts of blockchain innovation. A shrewd agreement, which is permanent on the grounds that its code is put away on the blockchain, allows for a variety of transactions to be carried out without the need for human involvement. Ethereum was the earliest to establish a brilliant agreement biological system, becoming the second most extensively used permission less blockchain crypto currency system in 2016. Shrewd agreement based utilizes range from decentralized casting a ballot to character check, and from worldwide cash moves to decentralized raising support, as per a few hopeful business visionaries and state run administrations. Blockchain innovation has a wide scope of utilizations since it takes into account disintermediation in regions like exchange, articulation, political interest, social association, and independence from the rat race. In any case, as we'll see later, there are issues and worries that should be analyzed and addressed for innovation to overall affect society.

7. Transparency

Defilement and basic liberties infringement regularly flourish in conditions of mystery, data imbalance and murky correspondence channels. As a conspicuous difference, blockchains are intended to get absolute straightforwardness to hubs the framework so each and every snippet of data can be followed to its source and finished ease. This makes casualties of misuse undetectable to society, making it challenging to help them successfully. There are numerous conceivable blockchain uses of straightforwardness. For instance, similarly as it is feasible to follow a specific decent from the maker to the buyer through blockchain-empowered production network applications, it is hypothetically conceivable to distinguish assuming there are missing people in light of the common data that exists on a wide organization of hubs associated with the equivalent blockchain or even interoperable blockchains. Given that moral and protection contemplations are considered, which are fundamental circumstances here to keep away from a situation of undesirable reconnaissance; it very well may be gainful to follow developments of exiles to work with knowing their whereabouts and giving them the assist them with requiring. Clearly, another condition connected with this is the capacity of displaced people to associate with the organization, which is a significant test. Concrete blockchain-empowered applications have been made to further develop straightforwardness and discernibility to help in battling illegal exploitation in the fish and precious stone mining businesses, handling youngster abuse and checking displaced people's characters utilizing applications introduced on their cell phones to empower them to get to fundamental administrations like schooling and medical services. In the realm of non-Bitcoin applications of delivered record technology, the concept of self-sovereignty has received the greatest attention. Although it has a lot of potential, it's important to remember that there are challenges present. including the degree to which individual data would be uncovered and perhaps liable to assaults [6]. To stay away from such assaults, endeavors like the DIACC/Secure Key exertion in Canada shows genuine guarantee of consolidating the advantages of the innovation with more recognizable outside trust secures for clients. One more significant case is the assurance of individuals' possession privileges of property, with land being one of the most feasible areas of use. There are various cases in numerous nations where land snatches have made gigantic misfortune numerous legal proprietors since possession archives can be fashioned. In a completely straightforward and commonly trusted blockchain-based public appropriated record of land proprietorship, it is preposterous to expect to swindle the framework and guarantee land possession without taking over management of the blockchain itself. This application has been on display in the Republic of Georgia since roughly 2016, and it is commonly cited as a convincing example of the idea that blockchain innovation can protect inhabitants' property. Sweden is also experimenting with blockchain-based clever agreements for land libraries. Fundamentally, it is possible to use blockchains to ensure impartial decisions because votes would be represented and decision- making fraud would be difficult but not impossible. A mid-2017 investigation by Nasdaq and e-Residency, a stage of electronic citizenship in Japan, came to the conclusion that an effective examination utilizing blockchain innovation in doing an e-casting a ballot interaction for investors. Blockchain innovation could likewise be utilized to identify defilement in government circles also, limit maltreatments of force in manners that conventional accounting techniques can't. By permitting writers and other public vested parties open admittance to public information on the blockchain, basic freedoms could be a significant recipient. The information could be utilized as certain proof to uncover criminal practices inside the state, and subsequently safeguard weak local area individuals. It is vital to take note of that the degree and execution of straightforwardness might shift starting with one blockchain then onto the next. It is possible to keep some information accessible to specific hubs while keeping the rest hidden in permissioned blockchains. For some businesses and administrations that rely on information transmission that is confidential, this may be of crucial importance. One of the blockchain development phases that enables the formation of permissioned blockchains in which the hubs can be configured with distinctive duties and authorization settings is the Hyper Ledger Fabric9. These platforms offer the opportunity to address a wide range of blockchain applications with varying degrees and levels of simplicity.

8. Blockchain Ethics - Conceptual Framework

This framework provides a precise outline for a careful investigation of important issues. The three tiers of blockchain impacts, as well as their ethical implications, are given here.

- First, the technological stack: any blockchain implementation and application is built on top of the underlying DLT stack. Within the technological sector, there are ethical difficulties with the technology stack.
- Second, there are different uses that provide different ethical challenges. We talk about digital currency, shrewd agreements (independently executable projects that empower programmed exchange and data handling), and business change to exhibit such application-explicit issues.
- Third, foundations and society: albeit the thought of decentralization is more extensive than blockchain - which is just a particular execution of conveyed record innovation - decentralization is a significant conceivable result of blockchain to directions and society.

The possible influence of decentralization on institutions and society raises broader and deeper ethical considerations. Decentralization is the essential advantage of blockchain at every one of the three levels previously illustrated. Therefore, the morals of blockchain can be better perceived considering the current conversation about the morals of distributed computing's centralization. People, partnerships, and legislatures have embraced distributed computing to set aside cash, and accordingly, they have become dependent on cloud specialist organizations. This pattern of centralization has clear cultural advantages. Notwithstanding, in light of the fact that it includes entrusting data to the cloud thus solidifying power in the possession of specialist co-ops, it presents moral worries. Two blockchain-explicit elements should be considered while examining the moral ramifications of blockchain decentralization. In any case, not all blockchain applications use decentralization. Second, there are advantages and downsides to blockchain decentralization. The moral ramifications of blockchain applications can't be evaluated without full information on these real factors.

9. Micro-Level Ethics Privacy

Security is a vital worry in innovation morals [7]. With the expansive reception of organization and distributed computing, it is normal practice to store information in unified cloud stages. Security concerns arise in regards to potential maltreatments of information to support the people who can covertly get to them without approval. In the blockchain innovation stack, the security of crude information decides the protection of clients and in this manner is central [8]. Major, general issues around protection incorporate what information would it be a good idea for one to uncover to others to partake in exchanges? What data ought to be kept from others and simply be noticeable to oneself? What are the appropriate circumstances and strategies for information sharing? It is in this manner fundamental to comprehend how the blockchain innovation stack can resolve these issues. In a blockchain, not all crude information are shared by all gatherings. The characters of members are shielded by e-wallet suppliers. The openly uncovered characters are wallet addresses. Assuming that an e-wallet supplier or a trade specialist co-op are gone after, the character of members would be uncovered and at serious risk. Blockchain is subsequently not altogether unknown. All exchange data is accessible to general society and this present reality personalities of members can in any case be followed through specific specialist co-ops. To address these worries, information sharing can be controlled in view of the necessities of the exchange. For digital currencies and other blockchain applications, it is feasible to deliver the theoretical of an exchange and offer it through the blockchain with all gatherings, while keeping the first exchange information in a customary data set. Blockchain isn't planned to supplant conventional information base. The information proprietors actually need the command over the information presented by conventional data sets. A cross breed approach that takes benefits of blockchain and regular innovation stacks to shape a common record is plausible. This approach safeguards fundamental confidential data as blockchain is used in getting and dividing data between parties. Blockchain requires all gatherings to endlessly share important information. All verifiable records are put away and can be gotten to all forever. This prompts a potential infringement to one side to be neglected. It is the pseudo-mysterious and super durable nature of blockchain that calls for advancements in security components. The circumstances and strategies for information sharing ought to rely upon the idea of exchange and the requirements of the information proprietor. An adaptable arrangement that fulfills information access control and worldwide validation is required. A creator of a blockchain application ought to constantly notice a data strategy that empowers exchanges and safeguards protection all the while.

9.1 Accuracy

At the point when lives and business results depend intensely on blockchain, information exactness is imperative for the dynamic cycles [5]. Information exactness requires the genuineness, devotion and precision of data [5], which blockchain addresses well. In the first place, the basic calculation guarantees that information and all verifiable records are by and large validated and acknowledged by all gatherings without brought together approval. This isn't just monetarily proficient yet in addition dependable. Second, the information are put away across the entire biological system, which can't be adjusted or mishandled. Invulnerability is guaranteed by the fundamental calculation, which doesn't depend on the accessibility of a reliable party. Blockchain is in this manner a characteristic, decentralized, and straightforward arrangement that maintains a strategic distance from the requirement for believe that generally emerges among trustors and legal administrators [9]. Nonetheless, blockchain can't ensure the first precision of all information shared on the blockchain. It can function as a trustless dispersed open record that guarantees that all information, once put away in the blockchain, are validated and unchanging. Any defective information created and brought by victimizers is integrated into authentic records assuming all gatherings concur upon it. All information are recorded and put away as long as the exchange is actually legitimate. In this way, precision should be guaranteed past the blockchain. Best case scenario, a start to finish arrangement in light of blockchain innovation ought to be completely implanted into the hidden innovation stack.

10. Meso-Level Ethics Payment Mechanisms

Due to the fact that cryptographic forms of money ensure cheaper and more effective instalments, businesses all over the world are starting to embrace digital forms of money, particularly Bitcoin. Many new businesses in the digital money ecosystem are currently offering cutting-edge technology and off-the-shelf business instalment arrangements, and the product and equipment of the innovation stack are becoming more accessible to end users. The fact that the entire connection is now unregulated puts this reception cycle under a moral test. A temporary lack of strict regulations provides excellent openings for the cryptographic forms of money ecosystems to proliferate through various business sectors at this early stage of their acceptance as payment methods. In addition, the lack of regulation allows criminals to use digital currencies as payment methods in mundane web transactions, evade paying taxes, and commit robbery while conducting business [3]. Critical concerns concerning the legality of cryptographic forms of money are raised by these mistreatments. Consequently, it is practically difficult to avoid controllers for a really long time, even though the style of thinking of digital currencies is against the rule. Further discussion of these moral concerns might result in practical recommendations for digital currency regulations.

10.1 Currency

Because cryptographic forms of money ensure cheaper and faster payments, businesses all over the world are starting to accept digital currencies, especially Bitcoin. The technology stack's hardware and software are becoming more accessible to end users, and numerous emerging businesses in the digital money sector now provide cutting-edge technology and ready-made business payment options. Nevertheless, because the entire cycle is currently unregulated, this reception contact confronts a moral test. In this early stage of the use of digital currencies as payment methods, a temporary rule that is drained of strictness provides beneficial openings for the biological systems that generate cryptographic forms of money. Simultaneously, the Cryptocurrencies, as advanced monetary standards, can assist with forestalling the utilization of fake cash and safeguard the freedoms of gatherings in financial exercises. Be that as it may, it is customary for national banks to control cash issuance to safeguard public financial sway firmly. Cryptographic forms of money are in direct clash with the laid out financial frameworks and, subsequently, state-upheld digital currencies might have a more promising time to come. Legislatures are additionally impacted on the grounds that the development of digital currencies unavoidably make moral difficulties for money related approach [10].

An unmistakable model is presented by the lawful store necessities forced on banks, which are fundamental instruments for risk the board. The size of such stores depends both on the liquidity of the store and the connection between cash market interest on the lookout. With the developing cryptographic forms of money market size, vulnerabilities around this relationship likewise develop and prompt difficulties in planning such save store prerequisites. Simultaneously, the viability of those lawful store necessities will decline, and monetary market chance will increment. An open market strategy is one more essential money related arrangement instrument often utilized by national banks to balance out the monetary inventory and invigorate financial development. The presentation of digital currencies definitely debilitates the predominant place of government issued currency and diminishes its influence on monetary development rates. Resources, for example, government protections are regularly exchanged government issued currency as opposed to digital currencies, however the course of digital forms of money to some degree replaces the government issued currency supply, making open market arrangements less effective in cash, financing cost, obligation, gold and security markets. The presence of digital currencies as an option money related supply loosens up the control that national banks have on the entrance of business banks and other monetary establishment to capital. This further debilitates the conventional money related approach of national banks. These impacts of digital currencies on financial frameworks are as a matter of fact significantly more convoluted. More methodical examinations are expected to resolve questions, for example, How to direct the half and half fiat-digital currencies monetary market? How to refresh existing financial approach instruments to target digital forms of money? How to exploit digital currencies for better monetary guideline rather than latently tolerating the presence of digital forms of money on the lookout? Further examination is expected to recognize and evaluate the impacts of digital forms of money on the viability of apparatuses utilized by the national banks. The reception cycle of digital forms of money shows network impacts, suggesting that without accomplishing the basic progress scale the advancement of digital currencies can be restricted. In this manner, the departure of digital forms of money is ruined by the enormous unpredictability of their costs comparative with government issued types of money, their unsure relationship with the financial framework, and the absence of guideline. Government support is crucial [12] and possibly compensating for states themselves. State run administrations and national banks might change guidelines and approaches to benefit instead of experience the ill effects of troublesome developments. A potential way for legislatures and national banks is to give the state-moved sovereign digital currencies notwithstanding customary government issued currency. The sovereign digital currencies would be controlled and upheld by sovereign ability to uphold their delicate status. The specialized instruments of the basic blockchain innovation stack would be equivalent to with other public digital currencies. In any case, all the figuring hubs should be licensed by the state run administrations. At last, however the total market worth of all digital forms of money is still generally little, their quickly developing exchanging costs raise concerns with respect to their worth. In this manner, there are worries about Bitcoin and other elective cryptographic forms of money also planned Ponzi Schemes. This discussion is basic and fundamental particularly when the crypto-resources are still under fast developing with additional adopters, less focus, multi-coin backing, and development [6]. The exploration gives an intensive conversation on this issue for additional examination. The lack of regulation allows criminals to use digital currencies as payment methods for shady web transactions, illicit tax evasion, tax avoidance, and shoplifting. Critical concerns concerning the legitimacy of digital forms of currency are raised by these mistreatments. Therefore, it is basically difficult to stay away from controllers for a really long time, regardless of whether the method that digital currencies think is against the rule. If these moral concerns are discussed further, acceptable regulations for digital currency might result.

11. Opportunities and Challenges Ahead Coin Mining

The giving of coins remains generally inside the excavator local area. Coin mining includes tackling testing numerical issues and in this way requires high power utilization. The benefits from mining exercises are corresponding to the limit of a digger's processing office. This interaction isn't biological agreeable and creates zero substantial worth. A mined coin is an element that exists in the computerized world in particular and has no presence in reality. Notwithstanding, the absence of purpose, worth and substantial quality of digital forms of money doesn't exclude them as monetary standards. The developing trouble of mining coins stays testing. It is generally simple to mine coins at first, however it later becomes testing on the grounds that the quantity of computations included increments over the long run. A rising number of estimations prompt higher mining costs and higher digital currencies costs in light of the fact that the expense of mining is subject to the expense of the energy consumed by mining. Extra bottlenecks emerge in view of the current specialized constraints including the confirmation interaction. Accordingly, coin costs might grow huge number of times far surpassing costs of government issued types of money. Expanding costs and mining costs infer that novices to the mining local area have inconveniences as early comers secure more coins and more worth at a lower cost. The rich get more extravagant and the strong become all the more remarkable in the excavator local area. These dishonest elements can be viewed as the "first sin" of digital forms of money. Besides, the costs of digital forms of money additionally intensely rely upon theory, which drive the emotional change of costs. This change prevents the usefulness of digital currencies as worth stockpiling and places the benefits of cryptographic forms of money in peril, exposing them to unforgiving analysis that qualifies them as shrewd. To save digital currencies from implosion, vital guideline and direction are normal.

12. Social Impacts of Blockchain Technology

The proficiency, straightforwardness, responsibility, and security related with blockchain enjoy produced significant benefits in giving computerized characters to the helpless, in giving admittance to microfinance to the unbanked, and in assistance Blockchain's straightforward, unchanging, and appropriated characteristics take out the requirement for outsiders, taking into account more proficient and practical exchanges. The blockchain's open-source nature permits anyone to take an interest, considering inclusivity while additionally decentralizing administration through the advancement of programming arrangements (public or private) customized to every nearby climate. As a result of the unchanging nature of this framework, no single individual can singularly change information on a chain, keeping up with security and probably limiting pay off and defilement. Anyway we have distinguished various difficulties and constraints. From an administration viewpoint, it is obvious that while utilizing blockchain innovation, decentralization and democratization of navigation are not ensured. The way where administration frameworks are formed is impacted by the kind of purpose. Customary go-betweens don't be guaranteed to see power relocating away from them, in view of the utilization cases examined in this article. As a general rule, they often utilize innovation to guarantee that they keep power. At the point when authority moves, it ordinarily shows up at new outsider entertainers instead of the innovation's end clients. In particular, from an advancement stance, the simple presence of innovation doesn't continuously suggest further developed incorporation, neediness decrease, or even most of the populace. The blockchain is a way to an objective, and the creators know about various use situations in which less expensive existing innovation would have been a superior fit. It becomes troublesome yet not difficult to incorporate the individuals who don't approach the web or shrewd cell phones in another advanced foundation. Indeed, even the presence of innovation won't have the effect it guarantees where human way of behaving isn't adjusted to an advanced life. The absence of meaningful organized exchange with agents from across the range of entertainers from the Global South dangers subverting blockchain-based apparatuses' groundbreaking potential. To comprehend the necessities of the networks that stand to benefit the most from this innovative change, significant exchange is required. It will likewise be important to lead itemized investigations of purpose cases to more readily comprehend the expenses and advantages of the blockchain in explicit applications; the entirety of the current writing and contextual analyses looked into for this paper highlight an absence of vigorous proof based research in the utilization of the blockchain to propel Agenda 2030. The absence of a considerable information base of illustrations learned for future applications is to some extent made sense of by the brief time frame length of involvement with this field, yet assuming we as a worldwide local area don't joke around about utilizing blockchain to address improvement and philanthropic difficulties, we should reinforce the exploration agenda [9]. Furthermore, it is important to acquire a more profound comprehension of what the execution of blockchain innovation might mean for administration frameworks. While blockchain innovation has been hailed for its decentralizing power, as the models above show, this might be more promotion than the real world. The utilization cases examined for this study show that the state (and, likewise, go-betweens who are expected to lose power) has been utilizing innovation to recentralize power. Accordingly, it's not obvious the way that the innovation is straightening administration frameworks, considering how the state and other outsider entertainers utilize the blockchain to engage themselves. Besides, the realities displayed here show that disintermediation is certainly not a characteristic aftereffect of blockchain, especially its 2.0 structure. Conventional outsiders, it shows up, are being supplanted with new ones: those with mechanical aptitude, coders, proprietors of enormous mining limit, etc. To all the more likely comprehend the ramifications of blockchain innovation on administration structures in the helpful and advancement fields, more examination is required. Nonetheless, in light of our primer investigation, there might be more frenzy encompassing the blockchain's decentralizing and democratizing propensities than meets the eye.

13. Tokenization in Social World Waste Management

A modern framework known as a "roundabout economy" is one that is beneficial or generative in both objective and plan. Through the inventive planning of resources, things, frameworks, and the presentation of fresh plans of action, the "finish of life" thought is replaced with the idea of rebuilding and advances the goal of waste end. Through various innovative advances, reusing has become progressively more practical through the arrangement of innovation like radio recurrence ID (RFID) and IoT. The arrangement of these advancements has made more prominent productivity regarding strategies, information sharing and following of materials. All the more as of late, different circularity drives have started utilizing blockchain innovation for check purposes. By reusing waste in the United Kingdom, RecycleToCoin aims to increase support. RecycleToCoin users are encouraged to bring their recyclable aluminium and plastic containers to participating retailers. Plastic Bank has made reusing frameworks in numerous worldwide areas offering fair benefit for squander gatherers and supporting computerized exchanges through the work of a permissioned blockchain. To accomplish adaptability and integrate different innovations, for example, investigation and visual acknowledgment, Plastic Bank has utilized blockchain on IBM Linux ONE. Food squander all through the world is a rising issue because of its effect on the climate, normal asset shortage and critical commitment to a dangerous atmospheric devation and environmental change. 33% of entire consumable is created is missing or squandered. The Sustainable Development Goals (SDGs) of the United Nations are expected to include an aim to reduce food losses in the production and supply chains, including post-reap losses, and to split per capita global food waste at the retail and consumer level by 2030. Food waste is a growing issue in Europe. The manufacturing, dispensation, and consumption of eatable usage common assets, and the elimination of food, especially is still palatable and the effects gets higher. To combat food insecurity and address the approximate USD 40 billion in tax advantages for organisations that go without being asked each year, one organisation in the United States is attempting to tackle garbage out of eatables. Goodr is leveraging blockchain to track an organization's excess garbage from eatables from gathering to gift, access charge., decreasing ozone depleting substance emanations and making social great by moving eatable excess to networks out of luck. Blockchain innovation can likewise be utilized in other waste administration ventures and assumes a crucial part in guaranteeing that recyclables don't wind up in landfill. Information tracking using computers takes into account more in-depth survey of supply chains. Incorporating this innovation with IoT devices along with RFID allows for the acquisition of more notable efficiencies and the underestimation of roque movement. Such behaviour has become typical in the global tyre recycling sector. One billion used tyres and 1.6 billion new tyres are produced annually, respectively. The collection, reuse, and removal of tyres from landfills, sensitive areas, and deserted areas continue to be challenges for the industry globally. The manufacturing of basic things, the manufacture of tyres, their dispensation, their application, and finally their collection, organisation, and further reuse make up the tyre inventory network (ELTs). Elastic grind is used with ELTs. There are consequently various partners engaged with the business, which incorporate unrefined substance makers, tire producers, purchasers and dealers, squander gatherers, recyclers, planned operations organizations and the public authority, who supervises the reusing business from an administrative and administration viewpoint. The basic building block from which a tyre is made in the production network is made up of natural materials. This serves as the tire's basic building block. This would allow the tyre manufacturers to tokenize the raw materials. As a token used to track raw materials within the inventory network, the token would refer to a computerised twin that represents natural ingredients. Due to this, partners can track the recently characterised resource across the entire stockpile chain. Digitalizing the raw materials in a tyre can be done in a number of different ways. The raw materials might be thought of as a non-fungible resource, a token-like resource that could be used to prove unquestionably who owns it and that it is legitimate. This could be done with the help of cryptography. Another choice is to have the resource characterized as a fungible resource; the resource can in this manner not be exchanged. Therefore, if a specific cluster of unrefined substance is anticipated to be followed across the shop network, at that point, this strategy would be more believable. The unrefined materials will be regarded as non-fungible in this instance. On this basis, the producers of raw substances create tokens, which are then given to the planned operating organisations. The strategic organisations transfer or boat the natural material and the same amount of tokens that are addressing how much unrefined component in weight to tyre production plant. The production facility gets the two tokens, organic materials, and tyres. The tyre manufacturer may continue using the token or symbolic language that refers to weight in this manner. Making a bunch token which addresses the mass of the basic things used to create a cluster of tyres is an additional option. Once manufactured, this group of tyres is transferred to a stockroom. The same amount of tokens that correspond to the clump of tyres are available for purchase from the tyre vendor, along with a bunch of tyres. In order to verify the legitimacy of the tyres in the store network as they are given off, a immediate reaction signal would be screened. Additionally, the tyre buyer can receive a sign that ensures the trustworthiness of the tyres they bought. As the tyres are substituted going forward and the ELTs are sent to garbage tyre mainframe, the entire system is fully perceptible and detectable.

14. Electrical Vehicles

The protection of data would be of utmost importance with several partner contributions, including numerous natural substance makers, tyre manufacturers, planned operations organisations, and reuse organisations. The secret issue of tokens using ZKPs could be used to regulate each partner's level of protection. To protect various partners, certain information might be kept confidential, and elements could be changed as needed to make them more perceptible. Responsibility for resources can be moved without uncovering the classified exchange subtleties, while as yet guaranteeing administrative consistence as per exceptionally characterized business rules. If an administration requires oversight of the production network, it very well may be given an examiner job using seeing keys to acquire industry experiences into the organization. Such experiences would include data on productivity, manufacturing yields, the refusal to sell used tyres again, and result validation from recyclers to determine handling allocations. With the use of blockchain technology, the arrangement of start to finish visibility into multiple venture store network networks enables various parties to monitor and follow finish-of-life tyres, building belief by documenting unquestionable exchanges on the designated record. By far the greatest proponent of electric vehicles, sales of electric vehicles in China increased from 1.2 million units in 2018 to north of 500 000 units. As of the end of 2018, there were 5.4 million light vehicle modules in use worldwide, with 3.2 million expected in 2019 [54]. It was tested to follow and control the Low Carbon Fuel Standard merit age through sun-powered chargers and electric-powered car chargers foundation, and merit swapping, using Power Ledger's blockchain-supported platform.

15. Autonomous Vehicles

By 2025, the trade is expected to touch 42 billion Dollars, making the rise of independent vehicles revolutionary. While moral difficulties regarding split-second navigation stay a main pressing issue, more prominent comprehension of the moral rules computerized reasoning (AI) will follow should be grasped in extraordinary profundity. While driving, people gain from their own errors. They seldom gain from others, aggregately misstepping the same way again and again. Artificial intelligence, then again, advances in an unexpected way. At the point when an independent vehicle makes a blunder, the other independent vehicles are all ready to gain from it. As a matter of fact, new independent vehicles will acquire the total range of abilities of their progenitors and companions; so by and large, these vehicles can learn quicker than individuals. It is inevitable that autonomous vehicles will circulate on roads shared by human-driven vehicles in the near future. Complex AI devices and distributed record innovation will encourage people to interact with others in ways that increase their likelihood of benefiting from their interactions. Independent automobile producers should make sure the invention is protected increases consumer incentives to build this trust and recognition. The social and machine worlds will eventually integrate. This will lead to imitation of the machine's common educational experiences by people.

The artificial hive will serve as a template to second people herd where we progress peacefully toward a future void of confusion and disasters.

16. Positive Examples of Blockchain for Social Impact

Blockchain is now being utilized by various privately owned businesses, states, and non-legislative associations (NGOs) to have a social effect. We should investigate a few contextual analyses and how blockchain could have extensive repercussions. As per a new examination, 40% of fish bought in cafés, commercial centers, and from fishmongers all over the planet was mislabeled and had hints of pig in specific cases. There was no production network that was straightforward. Think about what might occur if blockchain could follow boats, gets, markets, and conveyance. IBM has collaborated with various significant food organizations to utilize blockchain to make a straightforward inventory network for produce. Building brand trust is facilitated by people checking to see if the packaging and marks are accurate. If the food is nearby, people may follow each step of the production network and learn, for example, what pesticides were used and where it was grown. By including ongoing conveyance and instalment data, Agridigital, for instance, is enhancing grain supply chains between ranchers and markets.. Evidence Points utilized blockchain to move information from discernibility apparatuses across supply chains, permitting shoppers to check item claims of supportability or beginning. Due to blockchain, brands should demonstrate their positive effect guarantees. Clients can utilize their cash to pick fair and commendable providers in view of genuine facts [10]. The foundation and non-benefit industry could get a good deal on bank costs while additionally delivering money to help people in misfortune regions. Using the Ethereum blockchain, Consensys, a blockchain innovation company, created a money and voucher conspiracy for Oxfam on the Pacific island of Vanuatu. Help beneficiaries, shops, and Oxfam utilized blockchain and digital forms of money to lay out a framework that was open, fast, and straightforward, and was more affordable than banks. Blockchain has a wide scope of uses. Government spending, as well as monetary exchanges going from charge installments to profit charges, may be followed and public. On account of medical services, your clinical data and therapies may be safely kept up with and made promptly accessible to specialists in case of a crisis.

17. Perils of De-Centralisation

The web's advancement has instructed us that a decentralized and scattered organization might be vanquished by market influences and organizations. This ease of use could be supplied by proprietary software in the form of smartphone and computer apps that connect with different blockchain. Some of the world's largest IT firms have already begun to invest in blockchain technology. Furthermore, the reality of blockchain-based transactions has resulted in centralization systems that are vulnerable to failure. The way that digital currencies are exchanged on trades, for instance, seems to move the job of middle people from conventional banks and toward these trades. Not at all like banks, which are regularly safeguarded with thorough security, are trades liable to hacking and malware contamination, making them defenseless against possible robbery. One such episode happened in February 2014, when MtGox, the then-largest Bitcoin trade in the world, asserted that 850,000 Bitcoins having a place with shoppers had been taken. Information validness, straightforwardness, and productive sharing are completely guaranteed by blockchain [11]. For calculations and computerized reasoning to be moral, such straightforwardness is required. The quality and certainty of the information took care of into such calculations and man-made brainpower are improved by blockchain-empowered straightforwardness, tackling trash in trash out difficulties. Blockchain can work on the security of information and frameworks with regards to network safety, giving solid arrangements when dangers are consistent, the climate is perplexing, and conventional estimations are exorbitant. Besides, shrewd agreement applications are proper in an assortment of areas, including banking, exchanging, production network, protection, and administration. These applications have a wide scope of cultural impacts at different levels.

18. The Constraints of Obscurity

Individuals frequently imagine that Bitcoin or other block chain-based applications are mysterious. Indeed, even in permission less block chains that ordinarily don't need the arrangement of any ID or individual data to make a wallet and make exchanges, individual data may be uncovered when, for instance, the proprietor of cryptographic money needs to administer it for assistance or an item. Because of lawful and pragmatic necessities, giving individual data to those wishing to purchase administrations or items with digital currencies is frequently important. By then, the obscurity of the individual included would be endangered. Indeed, even on account of ATM machines that exist in a few nations as an approach to effortlessly buy Bitcoin utilizing money or MasterCard, there are a few careful steps taken by the ATM administrators to recognize the individual utilizing the help. On account of an ATM machine in Geneva, for instance, a Swiss cell phone number alongside private data are expected to purchase Bitcoins. By then, the wallet in which the Bitcoins are put away will be connected in the specialist co-op's data set to that specific individual. Also, CCTV cameras are supposed to be introduced at ATMs for security purposes. The primary special case for the absence of namelessness might be in the underground criminal world, frequently alluded to as the dark net, where exchanges can occur without connecting individual data to the person. This explains why the WannaCry ransomware developers asked for Bitcoin payments in instalments. However, any Bitcoin block explorer can track the addresses of those programmers, which might help law enforcement identify thieves who use those Bitcoins.

19. Government Opposition

Whenever the web originally arose, legislatures at first didn't give a lot of consideration. In any case, as it turned into an in a general sense significant power influencing exchange, media and correspondence, a few state run administrations began to oppose by forcing limitations and guidelines to restrict how the web was utilized. The equivalent might be expected for blockchain innovation, which is as yet in an early formative stage equivalent to the web of the mid-1990s. A few states are investigating blockchain and its applications so as to make government benefits more proficient, while others may not be so enthusiastic about the scope of monetary and social exercises possibly occurring on an unregulated and decentralized framework, as illustrated by the U.S. controllers' dismissal of a Bitcoin trade exchanged reserve. Specifically, the shared usefulness featured already will probably not be invited by legislatures who are submitting or authorizing human freedoms misuses or the individuals who wish to keep a hold on the monetary and data areas. It is significant, in any case, not to limit the real worries that numerous state run administrations may have with regards to potential purposes of digital money exchanges to launder cash, avoid duties and complete criminal operations in the bootleg market. While there have previously been recommendations for ways of alleviating such mishandles, the gamble stays genuine and could have a significant negative sway on social orders.

20. Administration and Rules in a World of Tokens

Tokenization might have an assortment of impacts on how associations are managed. Essential ideas in corporate administration, like responsibility, obligation, receptiveness, and trust, will be changed. Obviously, these associations' items and administrations will be affected, yet customary administrative models are fit for managing such disturbance. Considering the significant commotion to business brought about by tokenization, there is an arising assessment that current administrative types of corporate administration will battle and should be rethought. Tokens can be laid out to lay out a connection between the token and a monetary or non-financial environment, like customized information, and to record or move this information by means of PC code on a conveyed record. The digitization of all unmistakable and elusive resources, as well as the development of new sorts of privileges, makes huge administration concerns. It has been exhibited that by tokenizing genuine resources like protections, new corporate partners or token holders might have the option to impact the power balance inside organizations. Besides, as the dependence on mediators and outsiders diminishes, another kind of trust, computerized trust, in light of numerical calculations and machines, is being created. The quick advancement of appropriated records and blockchain builds the dependence on computerized trust. Therefore, how we draw in with different elements and society overall is affected by our computerized trust. Later on, individuals will have more confidence in decentralized frameworks of governance [12]. Blockchain innovation can tackle head specialist challenges that twist in the space of guideline and administration. People would have the option to confirm and control individual information access and use. Regulation would have to change to expressly distinguish information proprietorship and non-consensual information use. There is a compromise in the area of personal information between the instances in which private information ownership would further develop security assurance and control on an individual's information and the common people's want for information screening in example like working to grant scientists access to information for clinical consideration, general wellbeing, and exploration. People might be safeguarded because of expanded straightforwardness. Individuals will actually want to claim and safeguard individual information on the blockchain, conceding and cancelling access and permitting it to be utilized, shared, or erased depending on the situation. What information is accumulated and the way that it is handled is totally straightforward to every client. Disintermediation would make the most common way of endorsing admittance to individual information more straightforward, bringing about more effective activities. With the ascent in versatile application use, a bunch of freedoms is oftentimes conceded at the hour of download or join. While quitting is typically the best way to deny proceeded with admittance to information, the improvement of the current consent discourse in portable applications will bring about access control arrangements being put away on the blockchain, where the client will can alter or disavow admittance to recently conceded information.

21. Arising Solutions for the Protection of Human Data

Individual/advanced residents have not been the culprits of information misuse, but instead country state legislatures, security promoters, and common liberties associations. This reality brings up the issue of whether an individual is safe to information breaks including their own data. The level of seen or genuine abuse, e.g., individual ledger and money taken versus character extortion versus information spilling with no "immediate effect" because of a hierarchical break, may be the response to this speculative issue. Assuming the level of seen or genuine maltreatment surpasses "no immediate injury," the inquiry becomes whether the deplorable demonstration has made a singular shift from an uninvolved to a functioning computerized resident. Despite the fact that there is a ton of exploration on human hacking, there is very little in the method of observational and social investigation that shows a shift because of such maltreatment, basically at the full scale level [13]. Individuals (altogether) might be reluctant to convey the way that they have been casualties, so this question might be challenging to reply. Accordingly, the theory is that when the seriousness of the maltreatment is related, conduct change is possible. Nonetheless, trust can appear as a Leviathan a la Thomas Hobbes, where a focal power, for example, an administration association, capacities behind the scenes to guarantee that society observes the guidelines forced by the focal power. Distributed (P2P) trust is at the other limit of the scale, and it is constructed generally on associations and shared moral qualities. P2P trust can be relational, for example, among relatives or companions, or it tends to be laid out through a typical arrangement of functional guidelines, like the Inter-connected network. Moreover, the Internet Engineering Task Force, which creates conventions and designs to advance the Internet, utilizes "unpleasant agreement and running code" and "we reject rulers, presidents, and casting a ballot." Furthermore, notoriety diagrams can show distributed (P2P) trust connections. There are new arrangements not too far off that intend to forestall information releases and protect individual data. The European Union General Data Protection Regulation (GDPR) has laid out information insurance explicit to associations, including security by plan as an idea. Organizations would address a significant expense in the event that they don't follow GDPR. The GDPR additionally raises the issue of what establishes real information; for instance, hashing individual information on the blockchain could be risky. One idea was to cease from hashing actually recognizable data (PII). The utilization of zero information confirmation (ZKP) methods to forestall information spillage has been thought of, where reach values can be utilized rather than authentic human-individual advanced resident information. Such abilities are as of now being popularized. Also, there is a continuous discussion among scholastics and organizations on use cases and arrangements. Secure multiparty calculation (SMPC) is currently being utilized in the blockchain climate for protection authorized calculation and the on/off chain. This rundown of new arrangements is in no way, shape or form thorough, yet it shows rather, patterns toward computerized resident control of their information, as well as the complete use of human information protection there is still space for more examination to guarantee particular divulgence and control of advanced data.

22. Incosistent Approach

What makes blockchain innovation testing to convey and utilize successfully in some areas of the planet is the way that it requires quick and solid web access. The mining movement to confirm blockchain exchanges additionally requires huge handling power. Actually areas of the planet actually experience the ill effects of hugely powerless media transmission foundations [11]. Indeed, even created nations can have disparities that favor one cultural gathering or area [1]. The way that blockchains require circulation of information across different hubs compounds the issue because of the great transfer speed, handling power and capacity requests to be a dynamic hub. This separation would normally bring about having a few people, gatherings and districts less inclined to receive the rewards of the innovation for public administrations. Moreover, there is an information partition with regards to making and utilizing blockchains. Looking at the biggest blockchain organisations and emerging businesses around the world, it is clear that Western countries account for the majority of their locations [2]. In this way, the code and points of interaction created for the blockchain programming would likely favour particular communities and nations. The code would likewise be improved to conditions that expect a serious level of creativity, for example, high data transmission, stockpiling and handling limits, which are much of the time inaccessible in many emerging nations.

23. Trust in Code and Coders

There is always a community of core developers in the case of open-source software. While most blockchain do not require trusted parties to function, some do. There is a level of trust in the code and the individuals who work with it. This crew is enthusiastic and committed to improving the software. Patches should be applied as needed. The collaborative effort to identify and respond to threats as rapidly as possible. Trust has increased as a result of address patches made available through venues like the Bitcoin Wiki7. This did not stop some vulnerability from being exploited. In any case, for example, a blemish that brought about the production of north of 184 billion Bitcoins in an exchange that was shipped off two distinct addresses. The issue was immediately fixed. Indeed, even the new spotless rendition, nonetheless, has scaling limitations since it can deal with seven exchanges for each second [14]. This is a critical adaptability issue. If it has any desire to contend with other installment frameworks like VISA, which can have upwards of a huge number of clients. Upwards of 4000 exchanges each second are conceivable. SegWit and the Internet of Things (IoT) are two instances of new ventures. The Lightning Network should assist with this issue, yet there isn't one. The Bitcoin people group has come to a settlement on which choice to utilize. In any case, clients must choose the option to place their confidence in the engineers' choices. One more illustration of certain financial backers' appearing innocent confidence in blockchain innovation. The assault on the principal decentralization was completed by fans who put resources into the innovation. The hack exploited blemishes in brilliant agreement innovation, bringing about the deficiency of more than $60 million in Ether, Ethereum local cash [4]. Different burglaries in 2017 exploited defects in well-known programming utilized by the Ethereum people group, bringing about the deficiency of $34 million in Ether.

24. Disturbance Impacts

While many of the expectations and discussions surrounding blockchain technology focus on its exciting and productivity-boosting effects, there is also a belief that the next financial crisis may be caused by the rapidly growing and largely unrestrained and unregulated blockchain-based FinTech industry, which has increased the market cap of digital currencies to more than USD 140 billion as of Session. It is as of now deeply grounded that innovation, through computerization and robotisation, is bringing about the deficiency of numerous occupations that depend on low-gifted work. One could contend that blockchain innovation will drive robotisation to phenomenal levels, which would definitely prompt the further disposal of occupations in numerous new regions such as those depending on delegate organizations. The repercussions of wiping out the requirement for banks, insurance agency and, surprisingly, goliath online entertainment organizations, for instance, may have huge monetary ramifications for the workers of those organizations. This might have genuine negative ramifications for organizations that can't rapidly adjust to the approaching flood of blockchain advances. This maybe makes sense of the berserk and fast reception of blockchain advances by major worldwide IT companies and monetary organizations. History has instructed us that it is absurd to expect to keep mechanization advancements from supplanting position. All things being equal, significant changes in instructive and preparing programs need to be set up to permit the securing of the abilities expected to create, use and keep up with present and future blockchain-empowered frameworks that are supposed to supplant the go-betweens of today.

25. Prolonged Autocracy

The expanded interest in and utilization of appropriated record advances, for example, the blockchain should be grasped conversely, with the high grouping of client communications and information at the web's application layer, like on web crawlers, interpersonal organizations and content stages [13]. Supported by the ascent of cell phones and less expensive, simpler telecom access, the web has seen enormous development starting around 2000. Simultaneously, web clients have steadily gone to a moderately little arrangement of conglomerating stages to recover data and speak with each other [7]. It seems somewhat counterintuitive given that decentralisation was the foundation upon which the web was built that over time it turned out to be more cohesive. The fact that the wealth accumulated over the past ten or more years has primarily benefited a small number of international partnerships demonstrates this improvement, whose procedure was to connect web clients to one another by filling in as a client interface. Even though Uber is the largest cab company in the world, it has no vehicles. Facebook, the most well-known media company in the world, offers no satisfaction. The biggest store, Alibaba, is out of stock. Additionally, the largest provider of conveniences in the world, Airbnb, has no land [6]. Billions of individuals from all areas of the planet have become subject to the computerized items and administrations of few worldwide enterprises that serve, whether they need to or not, as stifle focuses for people's very own information, discourse and that's just the beginning [11]. While web clients remain exceptionally dispersed on a worldwide organization of organizations, the information these clients produce are a lot of moved in the possession of a couple of organizations. Because it takes less time and effort to get information or use other internet-based services, this interconnectedness benefits customers. Additionally, it enables a variety of independent businesses to thrive on these platforms and provides some online users, like video bloggers, with the opportunity to generate income by altering material using advertisements and other techniques [9]. But having such a high degree of centralization runs against to the web's original promise to be an open-source, interoperable organisation that can endure catastrophic losses without losing its core functionality [3]. Tim Berners-Lee, the inventor of the World Wide Web, has openly demanded that one of the very first ideas of decentralisation be revisited: Today's web is decentralised. The problem is that there is only one large social network, one web crawler, and one Twitter for microblogging. We have a social problem, not a problem with innovation. Recently, a few cyberattacks and targeted setbacks have demonstrated how powerless the unified model is to stop numerous abuses or breakdowns that could harm people, including the right to security and the freedom of expression. For instance, the Snowden revelations revealed that the American government has direct access to information from Google, Facebook, and other American businesses [5]. The 2012 LinkedIn hack, which resulted in the compromise of more than 100 million client accounts, serves as another example [2]. Information from Freedom House suggests that web restrictions, such as control, separating of content, and badgering of bloggers, to name a few instances, have been increasing since at least 2012, which is in favour of the chance for articulation [4]. It is important to keep in mind that in the middle of the 2000s, some attempts were made to lessen the severity of the centralization of online content services using P2P applications like Napster and BitTorrent conventions, which actually helped the centralization of power by utilising overlay organisations like Freenet, an amazing method for limiting monitoring and control [10]. However, using such tools also resulted in legal issues, particularly with regard to copyright infringements [7]. Several new developments focused more on limiting government and Internet service provider (ISP) reconnaissance. In order to achieve this, they avoided using businesses and, after giving it some thought, relied on various clients to act as middlemen. One such initiative is the Tor Project, one of the most aggressive open-source initiatives aimed towards creating a hindering safe online insight [13]. However, each of them still had limitations, such as not being able to reveal all of the organization's clients or preventing man-in- the-center attacks. Additionally, the early adoption and present uses of blockchain or other technical innovations cannot be understood solely by a response from weak against strong go-betweens. For example, BitTorrent was found to have been utilized to trade youngster erotic entertainment [8], Tor was credited with making commercial centers for illegal exploitation and different wrongdoings [6] and Bitcoin was the digital money utilized by programmers to blackmail assets from casualties as a trade-off for reestablishing admittance to important information [1]. All in all, remembering that is significant innovation isn't innately fortunate or unfortunate; rather it makes a bunch of possibilities that are informed by its center design and standards. Against this scenery, let us make a plunge in additional detail with a portion of the commitments and traps of blockchain innovation with regards to its effect on society at large.

26. Conclusion

This article examines the tremendous significance of blockchain decentralization, far and wide confusions about it, and its moral ramifications for administration, the economy, and society. At long last, new arrangement, innovation, application, and guideline (UTAR) standards for the moral reception of blockchain are offered, as well as an expected bearing for future is research on the morals of blockchain. Blockchain as of now has a wide scope of utilizations, and thus, it has expansive social implications. Notwithstanding, in light of the fact that the innovation establishment is as yet being created, these applications and implications are quickly developing. No single exploration can address all of square chain's moral issues since worries happen as the innovation creates and is embraced. To keep away from the negative ramifications of a game-changing innovation like blockchain, dire examination is required. This study will assist scholastics and specialists with turning out to be more mindful of and fathom the moral issues encompassing blockchain innovation. This report might actually fill in as a beginning stage for future blockchain morals research. The utilization of blockchain's problematic innovation to handle corporate, financial, administration, and social worries is a unique chance for humankind. To accomplish and understand blockchain's guaranteed decentralization at different levels, the unavoidable struggles and vulnerabilities should be perceived and managed with regards to morals and moral norms. All things considered, in the event that an innovation insurgency develops morally and ethically, it can understand helping the world potential. The blockchain is no exemption.

References

[1] https://www.mdpi.com/1999-5903/11/10/220/htm.
[2] https://www.emerald.com/insight/content/doi/10.1108/ITP-10-2018-0491/full/html.
[3] https://www.researchgate.net/publication/321012025_Blockchain_technology_for_social_imp act_opportunities_and_challenges_ahead.
[4] https://www.researchgate.net/publication/318131748_An_Overview_of_Blockchain_Technol ogy_Architecture_Consensus_and_Future_Trends.
[5] Hargrave, J., Sahdev, N. and Feldmeier, O. 2018. How value is created in tokenized assets. SSRN Electron. J. [Google Scholar] [CrossRef].

[6] Morlino, L. and Quaranta, M. 2016. What is the impact of the economic crisis on democracy? Evidence from Europe. Int. Political Sci. Rev., 37: 618–633. [Google Scholar] [CrossRef].
[7] Hargrave, J., Sahdev, N. and Feldmeier, O. 2018. How value is created in tokenized assets. SSRN Electron. J. [Google Scholar] [CrossRef].
[8] Hacker, P. 2019. Corporate Governance for Complex Cryptocurrencies? A Framework for Stability and Decision Making in Blockchain-Based Organizations; Forthcoming. *In*: Regulating Blockchain. Techno-Social and Legal Challenges; Oxford University Press: Oxford, UK. [Google Scholar].
[9] Sharma, M., Chapman, T. and Saran, S. 2018. A New Social Contract for the Digital Age; the Future of Work and Education for the Digital Age: T20 Argentina; CIPPEC: CABA, Argentina. [Google Scholar].
[10] Allen, D.W., Berg, C. and Novak, M. 2018. Blockchain: An entangled political economy approach. Journal of Public Finance and Public Choice, 33(2): 105–125.
[11] Atzori, M. 2017. Blockchain technology and decentralized governance: Is the state still necessary? Journal of Governance and Regulation, 6(1): 45–62.
[12] Böhme, R., Christin, N., Edelman, B. and Moore, T. 2015. Bitcoin: Economics, technology, and governance. Journal of Economic Perspectives, 29(2): 213–238.
[13] https://cordis.europa.eu/project/id/759681.

Chapter 2

Comparable Works in Blockchain Technology

Krishnapriya KS,[1] *M Rajeswari,*[2,*] *D Brindha,*[2] *C Sivamani,*[3] *M Sowmiya*[4] and *C Thilagavathi*[4]

Blockchain Technology plays a major role in today's environment and has numerous numbers of applications. Researchers are interested in applying Blockchain together with IOT, Data Science, Machine Learning and Deep Learning in healthcare sector, Agriculture, Industry applications, Education, Finance, Banking and many other sectors. The main purpose of adapting Blockchain or distributed ledger technology in research is that it provides higher security compared to other features such as flexibility and speed factor. It works based on the concept of decentralization and hence it eliminates the possibility of failure. It is desirable to use any of the four types of blockchain such as public blockchains, private blockchains, permissionless blockchains and permissioned blockchains according to the need. Another important property of blockchain is that it is immutable in which the records or transactions are not changeable. In blockchain technology, the

[1] Department of Computer Science, Valdosta State University, Valdosta, GA, USA.
[2] Department of CSE, Karunya Institute of Technology and Science, Coimbatore, India.
[3] Department of Biomedical Engineering, Kalaignar Karunanidhi Institute of Technology, Coimbatore.
[4] Department of IT, M. Kumarasamy College of Engineering, Karur, India.
Emails: kkottakkalsugath@valdosta.edu; brindha@karunya.edu; sivamani.chinnaswamy@gmail.com; sowmiya.m90@gmail.com; thilagavathic.it@mkce.ac.in
* Corresponding author: rajeswari@karunya.edu

transaction is transmitted over a peer-to-peer network whenever a new transaction arises. It validates the transaction for further processing as it is simple. This chapter discusses various applications of Blockchain technology in which how Internet of Things, Data Science and other sectors are interconnected to solve various real-time problems. The study initially concentrated on blockchain fundamentals, types, and design for IoT applications. Secondly, we discuss about energy efficiency, security, privacy, throughput, latency, block size, bandwidth, usability, multi-chain management, patch management, forks, independence, and discipline which are considered to be obstacles and issues associated with IoT applications. In conclusion, Blockchain's forth coming scope for the Internet of things are investigated. In accordance with the literature-based findings, Blockchain can be used to provide security and privacy solutions for IoT-related applications. In many fields, including health, finance, education, economics, and many others, the contemporary period is experiencing a significant technological change. The emergence of the Internet of Things is the primary cause of this transformation. The world has started to rely on far too many of these tactics that assist individuals in meeting their needs in the smallest amount of time and effort. Aside from that, the emergence of Blockchain technology contributes to this transformation. Worldwide, the Blockchain is revolutionizing financial and non-financial transactions. Individuals and institutions benefit significantly from these strategies.

1. Introduction

Humans have been residing in an information flood age. Information is multiplying at an extraordinary pace, and everybody is creating new knowledge every instant. "Data Analysis" describes transforming data into priority setting and intelligence work. It combines resources, technology, processes, and the methods humans and computers collaborate to Transform data into deep learning. Data science is suggested utilizing big data, and that is a field that turns data into valuable information [1].

Primary data originates from many sources and is categorized into formatted, unclassified, and semi-structured data. The majority of large data is unstructured or semi-structured, making data consumption a challenge. Information gathering, preservation, retrieval, collaboration, evaluation, administration, and visualization are all challenging at multiple levels, as is cybersecurity.

Knowledge management and blockchain-based may be used together to transform how we acquire and analyze information dramatically. The massive volume, heterogeneity, and quick rate of increase of knowledge and the continuous analysis of data technologies have created ultrahigh expectations on the management of user quantity, throughput, and clean energy of privacy rights requests. We begin with a review of data science, covering big data technologies and the risks it confronts. Although we adopt blockchain technology, we should enhance blockchain technology guidance and standards, focus on data security risk data and testing, continuously monitor development patterns, and aggressively explore legislative ways [1].

The unique properties of blockchain technology are exploited by a wide range of industries, including banking, medical, manufacturing, and education. In addition to being trustworthy, more collaborative, better organized, more credible, and more transparent, blockchain technology (BT) also allows for identification, credibility, and transparency. In this paper, we discuss some of the ways in which open science can benefit from this technology and its characteristics. The need for an open research ecosystem was identified to perform this. We compared them to BT properties to demonstrate that the technology is suitable for use as an infrastructure. We also examine relevant literature and prospective projects using blockchain technology for open science in order to provide a better understanding of the current research environment. The projects are classified based on their findings and relevance to open research, respectively. The major role that changes the view of technology is usage of blockchain in IoT [30].

2. IOT with Blockchain

By utilizing sensors and other network nodes and equipment, IoT is changing the way companies operate. IoT security organizations face a big challenge as a result of this change. The volume of connected devices has multiplied year after year, making information security progressively more challenging. IoT cybersecurity incidents can be prevented using blockchain technology in an IoT environment [32].

The combination of blockchain and the internet of things allows machine-to-machine communication. A shared ledger is distributed across all nodes that records a series of transactions stored in a database, validated by a number of sources, and published in a publicly accessible database. The integration of IoT with blockchain has several potential benefits, including the ability of an intelligent device to act freely without the need for central power. The system can also monitor the interaction between sensor nodes.

Despite Ethereum's architectural advantage, it could pose a challenge for IoT. Client-server and hub-and-spoke structures are common in IoT solutions. It is expected that this initiative will have a significant impact in the future, even though it is still in its early stages. The growth of the technology is supported by uniform cybersecurity norms and laws. By creating necessary rules for data receptiveness and mentoring connectivity, an eccentric level of protection to the network is offered by blockchain.

The IoT research sector is rapidly expanding, yet it is still sensitive to data security threats. IoT systems with resource limits and a scattered architecture cannot use traditional privacy-related measures. To address these challenges, experts are employing the blockchain system. Because of the development and popularization of blockchain technology, it is now being used in various sectors. *Blockchain* is a technology that has been on the rise for over a decade. Notwithstanding many breakthroughs in blockchain technology, IoT still has difficulties that require improvement.

IoT is a networked set of connected physical objects, machinery, processor architectures, humans or creatures, and artifacts that may transport datagrams without even humans as social contact. Security breaches via the network are possible throughout the connected mode; consequently, data transmission must be safe. Blockchain is a systematic collection of documents known as frames connected via encryption. Each block comprises the preceding block's encryption hash, a clock, and transaction information [2].

The blockchain is designed so that it is impervious to data alteration. It is a decentralized cryptocurrency that is available to the public. Blockchain can quickly capture transactions between parties continuously and systematically. Data cannot be changed retrospectively without affecting all following blocks when collected. The modification of the next block needs the network majority's approval. Other sectors can be improved in future work, such as the hash key created in the blockchain, and techniques may be built to handle this source of weakness. Even if authorization is offered in IoT networks, various vulnerabilities put data at risk; thus, techniques to tackle this problem may be announced soon [3].

Consumers and industries are adopting blockchain technology for various unique, diverse applications, from cryptocurrency to the food industry. As the number of these applications grows, so does their complexity. The amount of data held by Blockchains This data has been analyzed developed as a significant study area, resulting in improvements in information science methodology. In this case, Insights focuses on both upcoming research issues and exploring innovative applications ranging from Bitcoin price prediction to crime detection [3].

3. Recent Development on Blockchain

The Bitcoin cryptocurrency was the first Blockchain application. The emergence of Bitcoin has ushered in a new era known as the Blockchain. There are currently over 1000 Blockchain-based cryptocurrencies known as alt-coins. These advancements have piqued the general public's interest in Blockchain technology. Some analysts relate Blockchain's birth to introducing a double accounting system, which revolutionized the commercial sector. Voting (Social Krona), identification services (Bitnation, Hypr), provenance (Everledger, Chronicled), and copyright control are among the new Blockchain-based apps (LBRY, Blockphase). Although it is challenging to forecast Blockchain's future influence, it is reasonable to claim that it will enable a wide range of essential and diverse purpose [4].

Blockchain technology has lately sparked consumer and industry interest in various uses, including digital money, product safety, affordable healthcare, and weapon monitoring. As new Blockchain implementations emerge daily, the complexity and volume of data held by Blockchain grow fast, resulting in a new autonomous study path of Blockchain Data Science.

Blockchain design is the process of developing bitcoin, with the blockchain technology serving as a distributed record of cryptocurrency transactions in a digital, decentralized, trustworthy, and maintaining. The most well-known use of blockchain technology is bitcoin, built using a record of every transaction. Transactions enable hashing mechanisms to validate massive quantities of data. To create terabytes of data and ensure effective data analysis procedures, big data tasks need a vast quantity of computer space. The considerable influence on big data analytics necessitates a greater volume of data, and the data collected might vary by industry and company. Blockchains is a transaction ledger; the paper trail of transactions is shared in a trustworthy environment. The transactions and records are preserved and developing continually in a distributed way; anyone may see the transactions, but no single user has control over them. Data is saved in blocks linked together, and a hash function seals each block; they are centralized systems. Blockchain functions similarly to a public ledger; with various entries of transactions, new entries of transaction data are appended to the end of the block. If you modify the hash of any block, the hash of all following blocks also changes [6].

The article provides a synopsis of blockchain technology and large data platforms. Blockchain systems provide immutable, distributed, and decentralized data while emphasizing monetary and bitcoin systems. Extensive data systems focus on the importance of many enabling approaches in revolutionizing how data is processed and analyzed. More study is needed to include the many applications and use cases of blockchain technology with Big Data approaches [7].

Conventional gadgets become intelligent and autonomous under the Internet of Things (IoT) vision. This vision is becoming a reality owing to technological improvements, but there are still obstacles to overcome, notably in the security arena, such as data dependability. Given the expected expansion of the IoT in the following years, it is critical to offer trust in this massive incoming information stream. Blockchain has emerged as a critical technology that will change how we exchange information. Building trust in dispersed systems even without authority is a technical advancement that can transform numerous sectors, including the Internet of Things.

Innovations that disrupt its inception, IoT has used big data and cloud computing technologies to overcome its constraints, and Blockchain, we believe, will be one of the next. This study examines the issues in this connection and focuses on them. Blockchain IoT applications and analyses the most significant studies to examine why Blockchain may be used to enhance the Internet of Things [8].

Blockchain is predicted to transform the Internet of Things. The merging of these two technologies should be addressed while keeping the issues outlined in this study in mind. Adoption of legislation is critical to adopting Blockchain and IoT in government infrastructures. Individuals, governments, and businesses would benefit from this implementation. Furthermore, the agreement will contribute to the expansion of blockchains and the integration of IoT into mining processes. In practice, promoting embedded devices and maintaining data integrity may become a dichotomy.An examination of the viability of deploying blockchain nodes on Embedded applications has also been provided [9].

The Internet of Things (IoT) is highly significant in all fields. With the expansion of its capabilities, it is now employed in practically every industry. Because technology is utilized everywhere, essential security features are critical components that must be kept. A new technology known as Blockchain can be utilized to develop these principles. This technology is widely used to provide safe transactions between various goods without the involvement of a third party [11].

Throughout this article, IoT is infused with Blockchain to provide an overview of their major aspects, architecture, distinguishing properties, prospective solutions for various real-world situations, and alternative communication models. Despite the similarities between the two methods, both have certain limitations in each industry.

IoT, or the Internet of Things, is a technology that makes a wide variety of gadgets accessible via the internet. In conjunction with wireless communication and sensors, Radio Frequency Identification contributes to the growth of IoT Devices. Integrating controllers and electromechanical systems with intelligent features and IoT services platforms in order to develop interaction between cyberspace and the real world is how intelligent services are provided [10].

A variety of distinguishing properties of both technologies and their frameworks and applications are addressed. Because security is the primary issue in any model, numerous difficulties are utilized as guiding principles and open challenges. A new futuristic strategy may be developed using these open issues, resulting in a new analysis feature in these sorts of systems.

Since IoT networks are extremely large and scattered, security and privacy remain critical challenges. IoT devices that are resource constrained cannot use blockchain-based systems, which provide decentralized security and anonymity. They require high energy, latency, and processing overhead. The lightweight instantiation in our previous work was designed to be used in IoT and did not require proof of work (POW) and coin concept. The cloud storage, overlay, and bright house tiers of our strategy were demonstrated in an innovative home environment.

In every smart home, a high-resource device known as a "miner" is always on, coordinating all communication inside and outside the home. A miner also maintains a private and secure BC for managing and auditing communications. By extensively analyzing the security objectives of anonymity, consistency, and accessibility, our proposed BC-based intelligent home system is shown to be secure [13].

We were initially given a technique for addressing these issues by utilizing Bitcoin, a public blockchain of blocks. The concept was addressed using an intelligent house as a case study. We examined the numerous transactions and procedures related to the intelligent home layer as well as the many critical components of it. We also provided a comprehensive examination of its security and privacy. To our knowledge, this is the first study to optimize BC in the context of smart homes. Eventually, we will explore the applicability of our framework to more IoT areas [12].

As communication technologies continue to develop rapidly, the Internet of Things (IoT) plays a crucial role in its maturity and infancy. It has rapidly developed (grown) for significant data transfer via wireless communication. As a result, it is necessary to administer the system and meet the market requirements for practical use. Many present IoT systems have very centralized structures with numerous technological restrictions. Cyber assaults are one example of these constraints. As a result, new strategies for improving data access while protecting security and privacy are required. The solution to this challenge is to combine IoT with Blockchain, which guarantees sense-data integrity. The combination of IoT and Blockchain resulted in an immutable record that was extensive and detailed [15].

The Internet of Things (IoT) is growing year after year, with the ultimate objective of developing 5G technology. However, the expanded internet connection attracts potential assaults from various directions, making security and privacy difficult for IoT. Because the structure of IoT devices is decentralized, it is not easy to employ the standard current security approach used in IoT node communication.. It is treated as a distributed, decentralized, and publicly accessible shared register. Nonetheless, there are specific issues in the security arena when it comes to data dependability. As a result, there is an urgent requirement to provide confidence in a massive amount of incoming information [14].

Blockchain technology in this world, it has become one of the emerging technology among many businesses and scholars because it delivers excellent answers for the experiments related to the contemporary oriented architecture. Even though it is marketable or private, Blockchain is been distributed ledger that may guarantee the transformation of integrity where by reorganizing the ledger midst contributing users. On the other side, the Internet of Things (IoT) symbolizes an Internet revolution that can link practically all environmental equipment via the Internet to share data to generate innovative services and applications that improve our quality of life. Although the centralized IoT system delivers several benefits, it also introduces new obstacles. Such problems are possibly be resolved by merging IoT and blockchain technologies.

Next stage of development will be to combine IoT with blockchain technologies. Although centralized IoT design has many advantages, it poses high cost, scalability, and security issues. The Blockchain is a decentralized platform that can handle billions of actions between different IoT devices. As a result, the expenses associated with creating and operating big, centralized data centers will be reduced. Furthermore, in the absence of a third party, the security concerns associated with a single point of collapse will be omitted [17].

Blockchain technology has recently gained much interest. Even though it is still in the early stages of approval and faces several hurdles, it has the potential to be involved in practically every business. *Blockchain* is a tamper-proof distributed ledger that allows transactions to be carried out in a decentralized setting. It can resolve the major issues associated with the traditional centralized paradigm. On the other side, the Internet of Things (IoT) is been emerged as giant in the field of latest Internet growth [16].

The Internet of Things (IoT) is considered to be a critical source for transforming the components into innovative, which is composed of smart homes, smart cities, intelligent businesses. The Internet of Devices can connect billions of things simultaneously, intending to develop information exchange needs that improve our everyday lives [19].

Traditional equipment becomes intelligent and autonomous due to visualization of the various components on the Internet where by these matters become real and fact. The Blockchain looks to be an excellent security source for new-age technical applications. Because of the rapid rise of IoT-based innovative applications, there is a strong demand to safeguard the data from intrusion various factors like security integrity, and authentication of IoT-based applications are needed. The combination of these technologies may provide us with secure software solutions [18].

4. Security and Privacy Issues on Blockchain

Kunchang Li et al. [20] discussed the methodology which is mainly for set right to maintain the secrecy of data of minor particles and whereby they are into various problems such as related management of pseudonyms and real identities, lightweight signature authentication, safe, efficient data transmission and sharing. In order to attack above issues, need to implement block chain secrecy defense with very frothy and spreading the way out for minor grid intelligent valuable systems. Author has developed a unique algorithm for encryption and decryption and followed by the signature algorithm whereby pooled to safeguard the security and access resource on-demand. This system model includes entities, which are the component control the system like smart meters and aggregators. The designing goals are to be inserted which provide the double safety block chain with data accumulation and distribution process for satisfying the system model. Finally, they designed attack model which is the pricing media for the supply of power. In order to control the user's electric current data, the center is maintained with complete safety with maximum efficiency. To enhance the multi pricing of the control center in the system and protecting the secrecy data transfer which can shared with foreign users. It reflects that the output has performance with very low latency, high benefits by having very low response time with very effective cost which increase the demand in the secured system service field of smart grid. The system is split in to sequence of process which includes starting process, data development and accumulation process, and sharing process. Security analysis consider the correctness security based and perfection of the secure communication system with necessary encryption and intrusion detection by the signature algorithms used. This paper implements the advantage of the block chain technology in the development area of privacy defense with variety of group electricity data.

Ajit Muzumdar et al. [21], stated the present smart grid electricity purchase paradigm that faces issues such as privacy protection, openness and fixity of liveliness data, only way to destroy the data denial, non-benefited auction and also the poor buoyancy in the dispersal process of energy process. In order to evolve solution to the problems, the author offered a reliable and rewarded context based on spread ledger concept with and smart agreements, as well as various smart indentures such as energy instillation into insolent, grids, along with various energy like the bidding, energy trading, and usage. This conceptual model aids in the preservation of user privacy via pseudo identity, integrity of information and integrity via Spread-out ledger, against the context of poor performance through the distributed management data, improved marketing through vickery auction method is achieved by the elliptic curve based digital signature algorithm using the notion of non-repudiation. At the outset the overall structure of this work completely improves in order to innovate smart grid energy management, a safe and trustworthy framework for delivering energy based on demand. In a peer-to-peer network of users the system controlling contracts carry out energy exchange between them. Recorded the movement of the energy induced on a wide spread ledger technology, the Ethereum block chain framework, and using the PoS consensus process. The designed system includes four major modules: energy instillation, bidding, energy interchange, and energy ingesting.

Authors have set up an Ethereum blockchain network whereby planned to process a testbed at NIT Goa for performance validation, and this methodology is calculated by the interval of 15 and 30 minutes in the midst of the average time between the energy instillation, bidding and interchange process quantity with transactions and nodes sent with different range of rate (10 tps to 100 tps), whereby while carrying out 10000 trades of energy instillation and bidding of the entire system. *Winner percentage = Number of satisfied Total number of participants**100. This encourages more people to contribute the later part of the work energy transaction process, increasing use of renewable energy sources.

Nehaï et al. [22] elaborated the main characteristics of the block chain smart grid concept which in the midst of transaction do not exchange the energy which leads to the promotion of renewable energy in the locality and completely eradicate the fatalities of energy ih the system lines. The model is based on the connection of numerous Electric chains and to evade the hardship of the network it performs the circulate and standardize the hassles in electricity and. It is the tariff standards are recognized and an amount of energy produced to content the producer will profit of the maximum quantity of energy coin. In this survey, authors have considered three Blockchains are Bitcoin, Ethereum and SolarCoin. By using Bitcoin,it is declared to be one of the most system with the heaviest disadvantage remain the consensus protocol with the operation mode with greedy in energy. In Ethereum, Smart Contracts are characterized and it seems very useful. Finally, Solar Coin it is difficult to identify enough information about this technology. Later, knowing the variants of the Block chain and focused to target the growth of the block chain concept with the wonderful smart grid model.

Van Genderen [23] demonstrated all recent trends that cause the energy transition to be rapidly changing, and the electricity grid will face more and more challenges. In order to avoid these causing problems, it is convenient to have some application on top of the electricity grid. A smart grid is this application and delivered the efficient energy system with the heavy uses of intelligent transmission and spread-out of network. The main purpose of using a smart grid is the more balanced electricity demand profile, Islanding, Local cooperation, Market Integration. Now a days most people associate 'blockchain' directly with cryptocurrencies like Bitcoin. Even though this is the most known application, the technology of a blockchain can be used in many different ways. As well cryptocurrencies, blockchains are being experimented with in several start-ups in different fields of security, cloud storage, ticket purchasing and even contracts between doctor and patient. This gives an idea of smart grids and blockchains together with its security. An energy network is explained together with the energy transition and how smart grids are a solution to the current trends in the energy landscape. Applying a blockchain in a smart grid in order to keep track of all the transactions that are made, is currently not possible but maybe in the future it will. The main one is the energy it takes to add new blocks to the chain. Even with the most energy-friendly consensus mechanisms, the energy consumption is still significant. The blockchain uses two cryptographic concepts: a signature protocol and hash functions.

Xiaolin Fu at el. [31] stated that with the blistering- fire leads to the development of the thirst on the electricity enterprises the currently available environment problems with energy resources spreadout all over the rooftop photovoltaic solar panels. It also slide on the various touch of wind turbines and combination of heat and power which drags the global attention.The energy consumption become double since 2008, during the year 2017 with the effect of renewable energy causes 18 of the total electricity in U.S for the past 15 years of production.In this area the DERs activities are considered to be more eco friendly and economically real when it is associated with the archconservative energy whereby clues to the production of the natural energy which have the lower pollution with eminent reduction in the loss due to the transaction across the small grids in the system.While transforming to the distributed and consumer-interactive from the distributed and consumer-interactive it witnessed the current power transformation from a centralized, patron-centered network. To produce the fully reasonable appeal the Peer-to-Peer (P2P) energy supports the task of trading in micro grids, among DESs and EUs short of the third part involvement in the midst of the work. According to specific inclinations, the request actors are made to have conduct energy trading independently.The key note point in this work are two corridoe, the first one is the iterative double sale medium for P2P energy trading which obtain the good achievement of peak sharing of energy and the request symmetry. The independent consummation of this P2P energy trading model on blockchain. The attainment of the energy allocation with optimal rise and request equilibrium is made to reach the attained and justified. Along the development of the tool, they presented the social comparison results been calculated with optimal mean trade true cost value. To provide the guidance for the request actors, proposed the concept that in which the goods that are association of energy and extension of the request actor. The arrangement of the actual electricity requests has become the talk of the current forum, which affect the structure of the sale model's original request designs.

5. Electro Blocks a Block Chain-grounded Energy Trading Scheme for Smart Grid Systems

Sudeep Tanwar et al. [29] described that all over the world smart grid systems invite a demand response between the stoner and the service provider. But in utmost cases, they use centralized armature, during power generation which leads to a transmission loss and high overhead. Owing to that, there causes a mismatch between the two major factors like supplier and the end users. Also, here exists a peak plea for which there is high reliability in an automated energy exchange. Considering these consequences, they give a block chain grounded with an scheme of latest insolent energy dealing method, which provides a complete supportive and safeguarding energy exchange amid the stoner and the facility provider. Here, the block chain presents a non-public way to reuse deals through a translated the domestically happening sale interchange going to provide the solution with thoroughly eradicates the external customer for evaluation. Author has considered the study on various efficient usages of the solar panels and the shops which contains the original conduits and wide granges. Conservation of the data of Electro Blocks that whereby panhandler and dispatcher bumps their stabilities and storehouse quantities and existing sale on to the network. On comparing the performance of this mount-based system which is having the enactment of the central expert for the cost structure of the same type of parameters with the usefulness energy. They also proposed cost-apprehensive and store apprehensive algorithms and predictable achievement of the existing algorithms using realization criteria similar as detention, peripheral desires, and storehouse pullulate, storehouse prostration and stability. In contrast to store-apprehensive algorithms, cost-apprehensive algorithms are suited for druggies in terms of financial savings and a sustainable electrical network that eliminates the need to purchase and dispense from a central authority. The proposed Electro Blocks scheme would benefit from further exploration of its detention characteristics.

6. BlockDeepNet A Block Chain-grounded Secure Deep Learning for IoT Network

Among the various developments in IoT and 5G, Shailendra Rathore et al. [28] observed that big data is rapidly increasing in 5G due to rapid-fire development in both areas. DL has recently been considered a most promising approach for supporting big data analysis. Even so, there are certain challenges related to designing an effective IoT DL paradigm, such as single points of failure, sequestration leaks of precious data, and data poisoning attacks. In order to overcome these obstacles, they proposed a BlockDeepNet, a block chain-grounded combining DL with block chain to create a secure DL system. In the area of big data analysis in IoT, BlockDeepNet offers three advantages. It is essential to support Deep Learning at the IoT device location to alleviate sequestration leaks and to obtain enough data for Deep Learning. To ensure secure and reliable exchange of original and global updates, collaborative DL was deployed in a block chain environment. BlockDeepNets' effectiveness in real-time scripts was validated with a prototype model. As well as minimising challenges, BlockDeepNet achieves advanced delicacy with respectable quiescence and computation efficiency. The demand for further calculation power can be further impacted by DL at the device position. BlockDeepNet cannot accommodate this need. BlockDeepNet can be augmented with an unloading mechanism whereby bias with low calculation power can offload their DL tasks to the edge garcon via block chain transactions.

7. Block Chain for AI Review and Open Research Challenges

Artificial Intelligence (AI) and block chain technology were the most sought after eminent technologies in recent times, according to Salah et al. [27]. Using blockchain technology, one of the advantages of making a payment is the ability to ensure the payment is secure and reliable without involving an unbiased observer. Machines require intelligence and decision-making quality as important criteria. AI suggests intelligence and quality as essential criteria. Then, we bandied deeply about a check on exercising block chain technologies for AI. The block chain is a collection of information, which is arranged in a better way and is protected and connected with cryptographic attestations, and it consists of several collections of information that are arranged together. Three fresh characteristics come to the forefront of the discussion of block chain structures in contemporary infrastructures, where by spread all over that has embraced the spread out and participatory control, inflexible inspections that are regarded as random tests, and inherent skills. The structure provides both direct and indirect structure. The structure makes it possible for our customers to take advantage of the better qualitative models of variety of data when Block Chain is combined with Artificial Intelligence, which is a highly demanding field. As well as participating in the collection of training data, the models that are responsible for good coverage of the data collected are limited to the command of collecting the training data. Aside from the challenges posed by open exploration, there are foreseeable challenges caused by the integration of both technologies, such as the scalability and side chains of sequestration, the vulnerability of smart contracts, and the deterministic prosecution associated with smart contracts. A consensus protocol based on AI, a lack of norms and regulations, and the fog computing paradigm are all a result of AI-specific new consensus protocols.

8. IDS Based on Block Chains as well as Random Subspaces for SDN-enabled Industrial IoT Security

Nowadays, artificial control systems are being subjected to increasingly sophisticated cyberattacks that can have veritably detrimental effects on humans and their surroundings [26]. In order to deal with these issues, this paper concentrate on the security of commands in artificial IoT against forged commands and misrouting of commands. The block chain and the Software-defined network (SDN) technologies are combined together in the author proposed security armature. To protect in contradiction of the forged commands, the security armature which is the proposed system is framed with the interruption identification model with videcilet RSL-KNN, where by merge the Random Subspace Learning (RSL) and K-Nearest Neighbour (KNN). This supports the system misrouting attack which can help the misrouting attack, to some extent alters with the current rules of the SDN- which trigger the artificial IoT systems that focus the artificial control process and Block chain- grounded Integrity Checking System (BICS). Over the Industrial Control System Cyber-Attack Dataset the result of the mounted result was examined, which is an developmental area which group the software oriented network and block chain technologies which in turn provide overall valid yield. RSL-KNN leads to the fragile score of 91.07 below the double class and multi-class bracket activities autonomously. At the discovery of 100, BICS shall be elaborated the fraudulent inflow rules and with respect to the detection time whereby in scalable terms.For better IDS gracefulness it has been loaded with variety of literacy ways in this initial work, where in it can be examined by industrial control System (ICS) cyber-attacks datasets.It is found to be interesting to club the block chain into the fraudulent inflow rules available inside the inflow tables, than using it for only detecting them.

9. Conclusion

Blockchain technology is recognized as one among the emerging technology in the area of research today. Researchers started using Blockchain techniques together with other emerging methodologies such as IoT, Data Science, AI, Big Data Analytics and many. Blockchain techniques play a major role in wide variety of real-life scenarios such as banking, healthcare, retail, agriculture, education and so on. IoT has a great impact when it is applied in the field of Blockchain technology to build trust and rely security in IoT data. This chapter focuses on the great works carried out by various researchers in the filed of Blockchain technology. Researchers have contributed more and provided fruitful solutions for variety of issues.

References

[1] Lee Kuo Chuen, D. 2018. Handbook of digital currency. 2015. Elsevier Aujla G.S. et al., SecSVA: Secure storage, verification, and auditing of big data in the cloud environment. IEEE Communications Magazine, 56(1): 78–85.

[2] Ahmed Afif Monrat, Olov Schelén and Karl Andersson. 2019. A Survey of Blockchain From the Perspectives of Applications, Challenges, and Opportunitie. August 19, 2019, date of current version September 4, 2019.

[3] Alireza Esfahani, Georgios Mantas, Rainer Matischek, Firooz B. Saghezchi, Jonathan Rodriguez, Ani Bicaku, Silia Maksuti, Markus G. Tauber, Christoph Schmittner and Joaquim Bastos. 2014. A lightweight authentication mechanism for M2M communications in industrial IoT environment. IEEE INTERNE A. Marc, "Why bitcoins matters: https://dealbook.nytimes.com/2014/01/ 21/why-bitcoin-matters/," New York Times, vol. 21, 2014.

[4] Bartoletti, M. and Pompianu, L. 2017. An analysis of bitcoin op return metadata. In International Conference on Financial Cryptography and Data Security. Springer, pp. 218–230.

[5] Bano, S.A.S.-B. 2017. Sok: Consensus in the age of blockchains. arXiv preprint arXiv, 1711.
[6] Voshmgir, S. 2019. What is Blockchain? Retrieved July 2019 from https://blockchainhub.net/blockchain-intro/.
[7] Xiwei Xu, Ingo Weber and Mark Staples. 2019. Architecture for Blockchain Applications. Springer-Verlag.
[8] Díaz, M., Martín, C. and Rubio, B. 2016. State-of-the-art, challenges, and open issues in the integration of internet of things and cloud computing. Journal of Network and Computer Applications, 67: 99–117.
[9] Rivera, J. and van der Meulen, R. 2016. Forecast alert: Internet of things endpoints and associated services, Worldwide, 2016, Gartner.
[10] Nikoukar, A., Raza, S., Poole, A., Gunes, M. and Dezfouli, B. 2018. Low-power wireless for the internet of things: Standards and applications. IEEE Access, 6: 67893–67926. https://doi.org/ 10.1109/access.2018.2879189.
[11] Sisinni, E., Saifullah, A., Han, S., Jennehag, U. and Gidlund, M. 2018. Industrial internet of things: Challenges, opportunities, and directions. IEEE Transactions on Industrial Informatics, 14(11): 4724– 4734. https://doi.org/10.1109/tii.2018.2852491.
[12] Sicari, S., Rizzardi, A., Grieco, L.A. and Coen-Porisini, A. 2015. Security, privacy and trust in internet of things: The road ahead. Computer Networks, 76: 146–164.
[13] Roman, R., Zhou, J. and Lopez, J. 2013. On the features and challenges of security and privacy in distributed internet of things. Computer Networks, 57(10): 2266–2279.
[14] Dorri, A., Kanhere, S. and Jurdak, R. 2016. Blockchain in Internet of Things: Challenges and solutions. arXiv Preprint arXiv:1608.05187.
[15] Fraga-Lamas, P. 2016. Evolving military broadband wireless communication systems: WiMAX, LTE and WLAN. *In*: Proceedings of the International Conference on Military Communications and Information Systems (ICMCIS), 1–8.
[16] Stanciu, A. 2017. Blockchain based distributed control system for Edge Computing. *In*: 21st International Conference on Control Systems and Computer Science Blockchain, pp. 667–671.
[17] Lamport, L., Shostak, R. and Pease, M. 1982. The byzantine generals problem. ACM Trans. Program. Lang. Syst., 4(3): 382–401.
[18] Michael, J., Cohn, A. and Butcher, J.R. 2018. BlockChain technology. The Journal. Retrieved from: https://www.steptoe.com/images/content/1/7/v2/171967/LITFebMar18-Feature-Blockchain.pdf, 2018.
[19] Lin, J., Shen, Z. and Miao, C. 2017. Using blockchain technology to build trust in sharing LoRaWAN IoT. *In*: Proceedings of the 2nd International Conference on Crowd Science and Engineering, pp. 38–43.
[20] Kunchang Li, Yifan Yang, Shuhao Wang, Runhua Shi and Jianbin L. 2021. A lightweight privacy-preserving and sharing scheme with dual-blockchain for intelligent pricing system of smart grid. Computers and Security, 103: 102189.
[21] Ajit Muzumdar, Chirag Modi, Madhu, G.M. and Vyjayanthi, C. 2021. National institute of technology Goa, India. Journal of Network and Computer Applications, 183-184: 103074.
[22] Zeinab Nehaï and Guillaume Guérard. Integration of the Blockchain in a Smart Grid Model. https://www.researchgate.net/publication/320518549.
[23] Tim van Genderen. 2018. Blockchain in Smart Grids, June 29, 2018.
[24] Zhou, N.C., Liao, J.Q., Wang, Q.G., Li, C.Y. and Li, Y. 2019. Analysis and prospect of application status of deep learning in smart grid. Automat. Electr. Power Syst., 43(4): 180–197.
[25] Chen Zhang, Tao Yang and Yong Wang. 2021. Peer-to-Peer energy trading in a microgrid based on iterative double auction and blockchain. Sustainable Energy, Grids and Networks, 27: 100524.

[26] Abdelouahid Derhab, Mohamed Guerroumi, Abdu Gumaei, Leandros Maglaras, Mohamed Amine Ferrag, Mithun Mukherjee and Farrukh Aslam Khan. 2019. Blockchain and Random Subspace Learning-Based IDS for SDN-Enabled Industrial IoT Security, 15 July 2019.

[27] Salah, K., Rehman, M.H. and Nizamuddin, N. Blockchain for AI: Review and Open Research Challenges. 10.1109/ACCESS.2018.2890507, IEEE Access.

[28] Shailendra Rathore, Yi Pan and Jong Hyuk Park. 2019. BlockDeepNet: A Blockchain-Based Secure Deep Learning for IoT Network, 22 July 2019.

[29] Sudeep Tanwar, Shriya Kaneriya, Neeraj Kuma and Sherali Zeadally. 2020. ElectroBlocks: A blockchain-based energy trading scheme for smart grid systems, 25 June 2020.

[30] Nguyen, C.T., Hoang, D.T., Nguyen, D.N. Niyato, D., Nguyen, H.T. and Dutkiewicz, E. 2019. Proof-of-stake consensus mechanisms for future blockchain networks: fundamentals, applications and opportunities. In IEEE Access, 7: 85727–85745. doi: 10.1109/ACCESS.2019.2925010.

[31] Fu Xiaolin, Hong Wang and Zhijie Wang. 2020. Research on block-chain-based intelligent transaction and collaborative scheduling strategies for large grid. IEEE Access, 8: 151866–151877.

[32] Singh Rinki, Deepika Kukreja and Sharma, D. 2022. Blockchain-enabled access control to prevent cyber attacks in IoT: Systematic literature review. Frontiers in Big Data, 5.

Chapter 3

An Intergration of IoT and Blockchain for Medical Application

Dev Ananth S, Pranav Anand V, Kamalesh R, Jeevanantham M* and *Padmashree A*

Blockchain is a new discipline founded on the concept of a digitally distributed ledger and consensus algorithm this eliminates all risks associated with intermediaries. Its early applications were in the banking industry, but since then, it›s been expanded to almost everything major research areas, including education, banking, IoT, supply chain, governance, defence, and healthcare. Interoperability, security, authenticity, transparency, and simplified transactions and healthcare stakeholders made a demand on it by (research organizations, patients, providers, supply chain bearers and payers). By employing a patient-centric approach and eliminating the third party, blockchain technology, which is established over the internet, can utilize current data in healthcare in a peer-to-peer and that are interoperable one. Here we have produced a survey paper on block chain's significant role and how it operates in the healthcare sector.

Bannari Amman Institute of Technology, Sathyamangalam.

Emails: pranavanand.cb20@bitsathy.ac.in; KAMALESH.CB20@bitsathy.ac.in; jeevanantham.cb20@bitsathy.ac.in; padmashreea@bitsathy.ac.in

* Corresponding author: devananth.cb20@bitsathy.ac.in

1. Introduction

Medical care has grown increasingly important as the number of patients and diseases continues to rise. It is critical to keep track of a person's medical history to accurately handle any possible health concerns. The current climate ensures that health data is managed and shared clearly across diverse organizations. We don't know whether documents are securely exchanged with other organizations and there another strategy is given in the proposed work to avoid the effects of security and information trustworthiness. Maintaining the integrity of clinical records is critical. As a result, where there are attributes that are largely reliant on ascribes that fit, reports must be swapped. By digitizing and decentralizing medical care organizations, we can continue to expand the Responding to human wellbeing sector and observing patients in a clinical setting indefinitely and from afar. Produced clinical data is quite basic, and it must be handled with caution to avoid any tampering of data. In this situation, blockchain is an emerging technology that is the most widely accepted, decentralized stage. Without external management, it provides several powerful features such as sealed, changelessness, detectability, information reliability, classification, and security. Blockchain's feasibility for the medical services biological system has been identified by a few research papers. The studies [1–5] investigated existing research on using blockchain technology in medical care to improve security. Regardless, none of these studies have centered on the applications of Blockchain technologies and the (Internet of Medical Things) IMOT. In this unusual situation, we propose our study, which examines using efforts on blockchain reconciliation with the Medical IoT and talks about the specialized subtleties of each work.

1.1 Internet of Medical Thing

On each passing day, the number of related gadgets grows, and to keep up with the rapid addition of associated gadgets, a new organizational foundation is being organized and provided. With the Internet of Things, various creative firms have suggested a innovative vision of the Internet [6]. The Internet of Things (IoT) is rapidly expanding. It has become a massive component of the future. The IoT's main life pattern is divided into four sections: (1) to gather information using gadgets and sensors; (3) to save the gathered data in the cloud for investigation; (4) to send the dissected data back to the gadget; and (5) the device behaves in the same way. According to a Gartner report [7], By 2030, there will be 26 billion IoT devices, according to Cisco. According to indicators in the future years, a broad movement in connection between machines is expected which is ecpected to be used in variety of applications such as home mechanization, transportation [8], wearables [9], or heightened reality [10]. In the current state of remote communication, the IoT is grasping a lot of fame. IoT is defined as devices linked to the Internet that sources and admits all consistent data. An IoT device is fully electrical, from wearables to equipment gadgets, with a variety of uses and locations that are far away, The Medical IoT combines the advanced and real worlds to speed up the process of analysis and treatment, allowing doctors to focus more precisely on the patient's well-being and alter patient behavior and health condition in real-time. Patients and clinicians will be impacted significantly by the association of therapeutically related devices. The medical services industry is quickly integrating IoT-based arrangements. Medical IoT has also been given a forecast for 2022. By 2022, the associated clinical device of the Medical IoT for diagnosing, testing, and patient treatment are expected to grow from $14.9 billion–$52.2 billion [11]. To illustrate the Medical IoT concept, imagine an electronic clinical reports (EMR) being delivered from cclinical trial community to the patient's advanced cell or a wearable device For heart therapy, the patient wears an action tracker and physician monitor that on a PDA. These case show a basic machine-to-machine (M2M) communication that supports Medical IoT shown in Fig. 1.

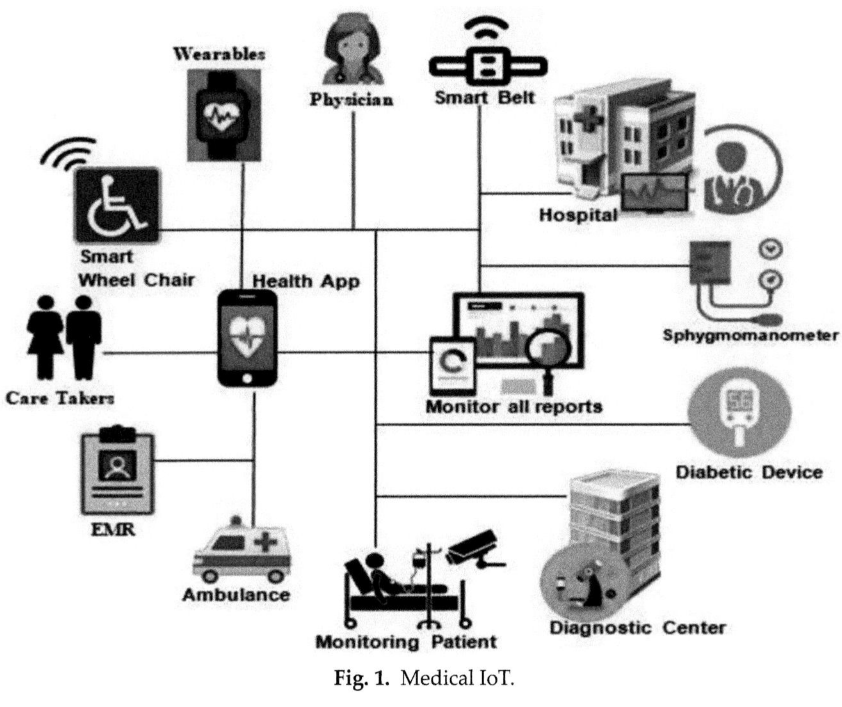

Fig. 1. Medical IoT.

1.2 Benefits and Positive Impact of IoT in Healthcare are Numerous and Varied

- Improving patient comfort and convenience leads to higher patient satisfaction and quicker healing periods.
- IoT (Internet of things) healthcare gadgets, wearable ones, and data accessors enable physicians to better monitor patients and give better therapy for patients.
- IoT security systems improve patient, physician, and staff safety.
- Sanitation systems that use UV light keep areas clean and prevent sickness.

Another benefit of IoT-savvy clinic setups is future-sealing. Even though there is an initial investment in setting up an IoT foundation, introducing devices, and training employees on programming frameworks, its value grows over time. When an IoT biological system is in place, it's simple to add new Medical IoT devices as they become available. A smart clinic powered by IoT will be a long time coming in front of comparative offices that decide not to carry out IoT gadgets and arrangements.

1.3 Challenges of IoT

Although there are numerous advantages to implementing IoT technology in healthcare, there are also numerous IoT problems. When it comes to deploying IoT in healthcare settings, healthcare facilities must be aware of the challenges they may face.

1.3.1 Security

Clinics and other clinical offices must adhere to stringent security standards, as well as patient protection requirements and consistency guidelines. Medical care organizations, particularly large medical care organizations, are frequently faced with overseeing security for a wide range of offices, as well as vast archives of data (Data collection is one of the most difficult IoT challenges in medical care, but it's also where the greatest opportunity for upgrade lies.) IoT frameworks generate a lot of data. It's vital to assess the framework's goal and needs for information collection and consumption before implementing an IoT environment in a medical services scenario. Obtaining IoT reports has become a major challenge due to advanced security challenges that have arisen as a result of the rapid growth and diverse nature of the industry During this time, previous security concerns have grown increasingly acute as a result, the security and protection of common IoT are at the forefront of our minds

[12]. Information security is important because it ensures dependability, authenticity, legitimacy, and information protection by storing and moving data with virtually no unapproved access, validated clients must be able to recover the protected data [13]. Because unknown clients are communicating with one another so cybercrime gradually harms devices and businesses. A great deal of data is acquired, transported, and communicated among distinct groups. The exchanges should be conducted securely as a result of this massive transformation, the path for digital attacks has become more open. Legitimate implementation of IoT made it at the physical level in IoT engineering so that any unapproved recipient can't access the framework. The Perception layer, Network layer, Middleware layer, Application layer, and Business layer are the five layers of IoT design [14]. Each layer has its objectives and problems. Confidentiality, Integrity, and Availability (CIA) are the three most important security objectives of the Internet of Things. In light of these flaws, there are four types of attacks in the Internet of Things: "Software attack," "Physical attack," "Network attack," and "Encryption attack." Even though Nexos core protocol, Power Over Ethernet (PoE), is one of the most secure technology solutions available, it's an excellent choice for healthcare facilities with strong security standards [15].

1.4 *An Overview of Blockchain Technology*

Satoshi Nakamoto first presented the blockchain data structure with Bitcoin in 2008. Nakamoto [16], creates an advanced record that allows for unalterable and irreversible exchanges. Blockchain is a widely used technology for exchanging and registering dispersed information. Even if they have no faith in one another, blockchain enables obscure groups to undertake numerous exchanges within the organization. Blockchain data structure trails and stocks data from a numerous devices devoid of using a central cloud. It's a digital ledger that adheres to an ever-expanding set of information standards. There is no unified technique and no expert computer to effectuate exchanges between hubs, this technology employs public-key cryptography. The exchanges are then recorded and placed on a public record. The record had a chain of squares that were cryptographically linked to one another if expecting to edit or remove or tamper squares of data that were previously documented on the blockchain is ludicrous. To update information to the blockchain, blockchain clients must solve a riddle known as a confirmation of work along with this, more consensus is introduced Practical Byzantine Fault Tolerance (PBFT), the Proof-of-work algorithm (PoW), Proof-of-stake calculation (PoS), and Delegated Proof Of Stake (DPoS) [17]. There are several essential concepts in the Blockchain such as the Previous Hash Code is one of them and each square must include the hash code associated with it, which serves as a unique identifier, the hash is created using a particularly puzzling hashing calculation. Each exchange's hash subtleties must be completed to be a portion of that square. The Merkle Root is a hex value that holds the exchange details of a square in the header. The value or proof of labour of that square is another essential idea related to the Blockchain; this is the numerical arrangement linked to the square to ensure that it is the main block. "Genesis block" is the absolute initial square each of the N blocks has a hash of the N-1 square. Figure 2 the exchange is visible to all members. The fact that the exchanges are visible does not imply that everyone will be able to view the real contact and the original contact is safeguarded with the help of the private key [18]. The blockchain utilization is exceeded the utilization of Bitcoin due to its varied highlights [19].

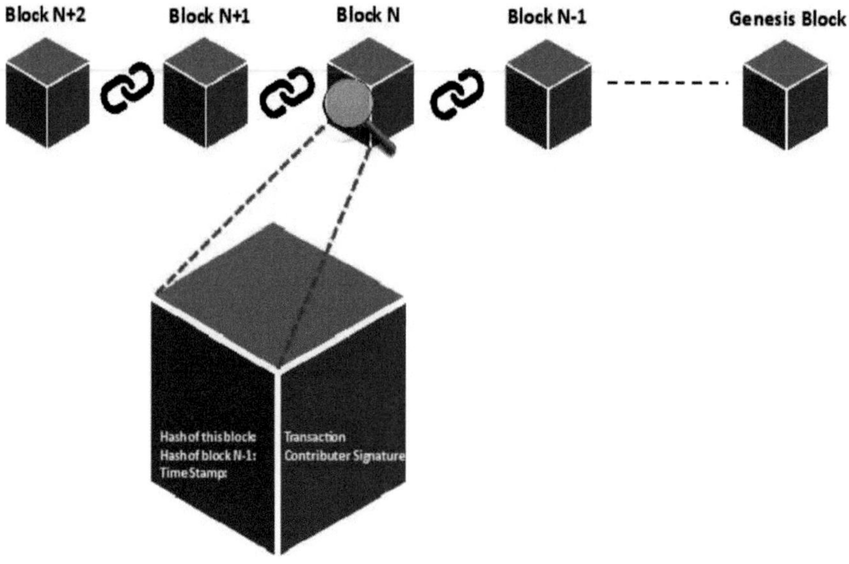

Fig. 2. Blockchain architecture.

1.4.1 *Types of Blockchain*

Blockchain has four types: private, public, consortium, and hyperledger [20].

1. Public blockchain

A public blockchain, by its very nature, is permissionless and entirely decentralized, allowing anybody to participate. All blockchain hubs have the same access to the blockchain, the capacity to generate new information squares, and the ability to approve new information squares with a public blockchain. Until now, the majority of public blockchain has been utilized for cryptographic currency trade and mining. Well-known public blockchain like Bitcoin, Ethereum, and Litecoin may be familiar to you. The hubs "mine" for digital money on this public blockchain by addressing cryptographic limitations and creating blocks for the exchangers listed on the platform as a reward for their hard work; these excavator hubs are given a limited amount of cryptographic money. The excavators are essentially new-generation bank employees who are planning a transaction and need to get (or "mine") a cost for their efforts.

2. Private blockchain (managed chain)

Private Blockchain or the monitored blockchain that is a permissioned blockchain controlled by a single entity. Here, the pivotal authority selects the hub, when it comes to filling positions, the central authority does not give each hub the same level of authority. Since the blockchain community is small, the private blockchain is only partly distributed. A few applications like Hyperledger and Ripple fall under private blockchain. These are used in cash trading that is done virtually in businesses. Private blockchain is more subject to falsification and agitators. To address these flaws, a consortium was formed, with a half and half blockchain created.

3. Consortium blockchain

Consortium blockchain, unlike private blockchain, are permissioned blockchains that are represented rather than a single company, by a group of organisations. As a result, consortium blockchain has a higher level of decentralization than private blockchain, resulting in increased security. Consortia formation, on other hand, can be a full interaction because it demands participation from multiple organizations, which brings calculated challenges as well as the potential for antitrust ruins. Additionally, a few supply chain actors might deficit the essential inventiveness or foundation to deploy the tools, while those that do may feel that the upfront expenditures of digitizing the information and interacting with new supply chain participants are high a charge to pay. R3, a venture programming firm, is noted for its consortium financial administrations that can benefit from blockchain solutions . CargoSmart has formed the Global Transport Trade Setup Association, a non-profit blockchain collaboration to digitize the delivery business and make it easier for maritime sector administrators to interact [21].

4. Hybrid blockchain

The term "hybrid blockchain" refers to a particular kind of blockchain that combines elements of both public and private blockchains or aims to utilise the best of both. In a hybrid blockchain, transactions and data are kept secret, but they can be validated when necessary by granting access through a smart contract. Private information is saved on the network, but it may still be verified. A private entity cannot alter transactions on the hybrid blockchain, even if it owns it. Businesses can build up both a private, permission-based system and a public, permissionless system using a hybrid blockchain, allowing them to control which data is accessible on the blockchain and which data will be made public.

1.4.2 Blockchain and IoT Technology – An Intergration

The number of patients is steadily increasing, and it is becoming increasingly difficult to provide comprehensive clinical care. With the use of IoT and wearable gadgets, the character of clinical consideration has improved in the last few years [22]. The fundamental mechanism for addressing medical care difficulties is far away constant observation. Wearable devices are used to collect and transmit data to emergency rooms, and in remote patient care, IoT devices play a critical role [23]. The main purpose of those devices is to provide vital relational data to healthcare providers, such as a person's breathing patterns, blood glucose level, and pulse [24]. Medical information-gathering devices are divided into four categories: Medical Embedded devices are implanted within the body of patients, Medical wearable devices are advised by physicians, and Wearable Health Observing Devices that are adviced to wear. RPM's fundamental thought process is to obtain information using blockchain technology. By adopting the concept of decentralization, block tie facilitates obtaining information from multiple cyberattacks. Blockchain also uses clever agreements to verify the information. Medical care is an IoT framework application that necessitates certain supplementary requirements such as interoperability and data transfer to get the job done information about the patient; the term "interoperability" refers to the most prevalent method of providing data to many sources. The idea of centralism covers a constraint to reach interoperability. The Internet of Things is depends upon centralizated one, in which data are stored in cloud that can't be accessible. Medical care apps face security challenges that blockchain combined with IoT can solve [25]. In the case of medical services, many tests have already been conducted in Blockchain [26].

1.4.3 Ethereum-based Contributions

A private Ethereum-based approach to make clever arrangements to manage client/device requests and control access [27]. To protect spectacular arrangements from hazardous activities, the creators offered a Product Ownership Management System (PoMS) as an alternative to the PoS arrangement concept. PoMS licenses collaborators with a large percentage of clinical data known as tokens with support and make blocks. A perspective of a variety of properties, including authorizations, work, and the surrounding area. IPFS is used to set a data limit. Interplanetary File System (IPFS) is used to store patient well-being records and device-specific data.

Makers presented a permissioned blockchain-based approach for safe far-away tolerant checking in [28]. They employed Ethereum to carry out devious plans to review data and convey warnings to patients and clinical care providers. Instead of using the PoW understanding model, they suggested using Practical Byzantine Fault Tolerance (PBFT). The proposed plan requires ways to handle the issues of coordinating IoT and blockchain. Furthermore, magnificent arrangements are applied to government permission to data in SMEAD [18], an Ethereum-based plan for detecting diabetes sufferers.

1.4.4 Modified Consensus Protocol

A few words, such as [28], have advocated adjusting the agreement convention to match the clinical IoT specificities. In [29], the authors presented a consortium blockchain-based engineering to securely record data generated by IoT while ensuring the patient's protection. Patient specialist programming (PA) is used in the proposed engineering to characterize the Blockchain's capabilities. It's constructed on a cloud server with carefully built capacity for the huge volume of health data, and it's supplied on an Edge processing organization for lightweight activities. The developers also proposed a modified PoS agreement, which includes appointing a pioneer to approve and build squares for a set of hubs. To handle health information, such as sorting clinically worthless health data and generating new agreements and brilliant agreements are utilized. Keep an eye out for specific events, and if necessary, migrate data to the cloud, characterize data, and so on. The updated PoS, according to the inventors, is more energy-efficient and has a shorter square age time than PoS.

1.4.5 Modified Cryptographic Technique

Authors developed a modified blockchain-based topology for clinical IoT devices in [30]. To begin with, the introduced blockchain is private:

hubs must be certificated to enter into the organization and send the transactions. Second, the POW convention was abolished b0079;'0 the founders. The group scrambled information in squares and stored all the interconnected squares in the cloud to deal with the enormous volume produced by Medical IoT equipment. To ensure sealed capacity, the hashes of squares are recorded on the blockchain. They use a 'lightweight protection safeguarding ring mark conspiracy' for anonymity and the authenticity of the client, This enables a set of hubs to take part in the data signature. Obtaining and guaranteeing the accuracy of data throughout the transmission and capacity the developers utilized a two-fold encryption approach in addition to the sophisticated mark. The information is scrambled using lightweight ARX calculations, and the key is encoded using the beneficiary's public key. The creators devised the Diffie-Hellmman key-trading method to make the exchange of public keys more convenient. Hubs are grouped to provide adaptability and organization while putting off problems. A bunch head is chosen to confirm and store hash blocks, check advanced marks, and keep track of communications among the group's hubs. The requested job is not completed, and no assessment is made. The writers of [29] continued to work together although their agreement convention had altered to improve patient safety, the ring signature was recommended as an alternative to the usual public key-based advanced mark.

1.4.6 Hyperledger-based Contributions

In [31], the researchers introduced IoT-blockchain-based engineering for remote medical care monitoring. In the field of engineering, there are two types of blockchain: (1) Medical Equipment (2) Blockchain-based consultation for storing clinical data generated by medical devices during therapy. Patients' records are kept indefinitely by emergency clinics using blockchain. Fabric's smart contracts (Chaincodes), which are conducted by embracing peers using a technique called Practical Byzantine Fault Tolerance, are used to verify and approve the exchanges. The designers created a user interface that allows patients' health data to be visualized.

In information transmission between networks, getting data is a critical problem. Medical IoT apps and stages that rely on a single cloud are jeopardizing security and our goal is to create a secure approach or instrument that provides privacy and validity with blockchain innovation in Medical IoT, as well as the integrity of information transmission.

The traditional system of the cloud compromises the security [32], however, it is a safe and unavoidable digital record that no one can change, increasing the precision of records. In cloud computing, the information are in concentrated structure, whereas in blockchain the information are placed in decentralized structure without any central hub and no outsider access. In terms of security and safety, blockchain outperforms the cloud in terms of audibility and transparency of transmitted data. Blockchain provides shared exchange with no delegate, and each hub adjusts the exchange rather than a central point.

1.4.7 Types of Attacks

1.4.7.1 Physical Attack

(i) Node tempering: To obtain the encryption key, Hacker changes the compromised node.
(ii) Physical harm: The attacker causes physical harm to IoT system components, which leads to a DoS attack.
(iii) Malicious code injunction: An exploit called malicious code injection provides an attacker total control over an (INTERNET OF THINGS) IoT system.
(iv) RF Interference on RFID: The offender interferes with RFID transmission by sending noise signals across radio frequency signals.
(v) Social engineering: To achieve his goals, the attacker obtains private information from an IoT system user.

In a sleep deprivation attack, the attacker's main objective is to bring down nodes. An attacker can stop wireless communication by using a jammer in a node jamming assault on wireless sensor networks.

1.4.7.2 Software Attack

Phishing attacks happen frequently. The attacker uses phony websites to gather the user's private information. Viruses, worms, Trojan horses, spyware, and adware are malicious programs that can spread over email attachments and the Internet. Worms can replicate themselves without human intervention. Scripts that are malicious: This exploit gives users access to the system. DOS: The adversary's main goal is blocking the functions of users.

1.4.7.3 Network Attack

(i) Attacks involving traffic analysis involve intercepting and studying messages in order to collect network data. (ii) RFID spoofing: When an Hacker modifies RFID signals, the system receives misleading information.

(iii) The system accepts the attacker's modification of the information. Sinkhole attacks are a common type of assault; their main objective is to deceive surrounding nodes into thinking they are on a specific route. (iv) The Sybil attack allows a malicious node to take on the identities of multiple other nodes by inserting a malicious node into the network.

1.4.7.4 Encryption Attack

(i) This attack's main objective is to get the secret key, which is necessary for two devices to communicate. When a communication is transmitted from the user to the server or vice versa in this attack, the attackers publishes more valuable datas. (ii) An attacks utilizing cryptanalysis: in these attacks, the attacker changes a message's format from one that is unintelligible to one that is, with or without the use of a key, understandable. (iii) Man-in-the-middle attack: To steal sensitive data, the attacker continuously monitors the nodes' communications.

There are numerous security tips in the literature. However, remaining issues like centralization, single points of failure, etc. make security in IoT networks a concern. In order to increase IoT security, a cutting-edge technology called Blockchain can be deployed. The potential of blockchain technology may be applied in IoT to enhance its safety and make it a safer secure network by solving the challenges and issues of centralization in the present security processes and introducing the notion of decentralisation employing the blockchain.

A blockchain is a decentralised point-to-point network that eliminates the need for an intermediary and enables direct communication and transaction execution. Each transaction is unique from the others and executes independently. Blockchain technology underpins the well-known and ground-breaking idea of cryptocurrencies. Cryptocurrency is said to be extremely safe and impossible to hack. The enhanced security provided by the same Blockchain technology may be used by other networks. Through the blockchain, everyone has access to a public distributed ledger system. A blockchain is a shared database of documents that keeps information in time order. Information regarding a transaction is included in a block. Each block is made up of data and hash of a block before it, and the relevant block's hash. The entire document is composed of the Transaction and Header sections. Details about the blocks is contained in the header. The moment the block was formed is recorded by "Timestamp." The "difficulty level" of a block determines how challenging it will be to mine. The fingerprints of each transaction in a block are represented by "Merkle Root," and the answer to the mathematical conundrum posed by the Proof-of-work algorithm is "NONCE."

1.4.8 Blockchain Technology and Related Concepts

Distributed ledger technology a new that is being utilized in many different networks to guarantee their dependability and security. Additionally preferred in many transaction management systems, the current transaction management system is being replaced in Blockchain technology.

The following are some problems with the existing financial system:

(i) High transaction costs,
(ii) Double spending, and
(iii) Cries are now associated with banks.

Blockchain, the main technology underlying bitcoin, has addressed the problem caused by centralized banks with its decentralized feature. The blockchain, a distributed, open database, houses the encrypted ledger. A central coordinating system exists in a centralised design, and each node is linked to it. All information between the nodes will be communicated, transmitted, and authorised through this central coordinating mechanism. Under this approach, each of these independent dependent nodes will be disconnected if the central coordinating platform fails. Therefore, it is essential that we switch from a centralised to a decentralised system. More than one coordinator will be present in the decentralised system.

There is no centralized authority, in other words. Since every node is connected to another node, there is no need for a specific coordinator in this system.

Each block in the chain of blocks that makes up the blockchain is a compilation of all recently completed, validated transactions. The sequence of blocks and the overall structure of the Blockchain are depicted together with the crypto graphical connections between each block. Each block contains all of these transactional characteristics as well as a block-by-block computation and storage of a consolidated hash code. As soon as the transaction is validated, this block is added permanently to the blockchain, and the chain continues to extend.

The second most advanced technology after the well-known bitcoin is blockchain. Understanding how bitcoins operate on the Blockchain can improve understanding of the technology. Santoshi Nakamoto released the first decentralised digital money, Bitcoin, in 2009. Bitcoin's generation and management are controlled and secure thanks to a variety of cryptographic techniques and mathematical ideas. The cryptography and algorithmic technologies are continuously updated. The Blockchain is the name of the electronic, highly secure ledger that records the total amount of bitcoin transactions.

A number of essential principles are present in the Blockchain. The Previous Hash Code is one of them. Each block must have the hash code that serves as an identification element for that block. The hashing algorithm used to produce this hash is extremely complicated. To be a part of that block, the hash information for every transaction that has occurred must be complete. The Merkle Root hex value in the block header contains the transaction information, the worth or evidence of the block's labor is another crucial idea related to the blockchain. To confirm that this is the correct block, the block's mathematical solution is attached to further understand how the Blockchain functions, let's look at an example. Let's say A wishes to send B money. Every node in the network receives a broadcast of the block that serves as the representation of the transaction. After that, the transaction can be approved if there is a large enough group of miners. The transaction is posted to the Blockchain after receiving approval from miner who successfully complete the proof of concepts, and B subsequently receives the funds.

The most important component of a Blockchain is a block, which is a permanent database that records all of the most recent transactions. Three technologies combine to form the blockchain. The Blockchain is firstly made immutable by using hash function and private key encryption to secure identities. It utilises a peer-to-peer (P2P) network that guarantees total consistency with the Blockchain.

Since lot of network users still have access to the old blockchain and won't accept the updated block, if someone attempts to make a little modification to a transaction and block that forms a component of a blockchain, the new block won't be added to or reflected in the Blockchain.

The software used to generate the Blockchain contains numerous security measures and protocols. The most popular language for creating Blockchain programmes is Solidity. Every transaction that is checked and authenticated in the development of a new block in a Blockchain is recorded together with details about the time, date, participants, and amount that is sent over. The entire Blockchain is held by each person that is a participant in it. Each transaction in the Blockchain is validated by the miner after they have solved a challenging mathematical challenge. Once the puzzle has been solved, the transaction is verified and kept in the ledger.

1.4.8.1 Public Distributed Ledger

Everyone has access to the data contained in a blockchain. With this, you may see the complete history of transactions that have occurred since the Blockchain was founded as long as you are a member of the network.

The user must consent before any changes may be made to Blockchain. Any changes to the Blockchain must be approved by a majority of the network's participants. This is the section of the ledger that is "public". One way to think of Hyperledger is as a piece of software that anyone can use to build their own unique Blockchain service. Only parties that are intimately involved in the transaction are updated and alerted on the Hyperledger network.

1.4.8.2 Hashing Encryption

Hashing encryption is used by the Blockchain to maintain security. The hash function is used in blockchain to carry out cryptography. The header's Merkle Root hexadecimal value holds the transactional information. Blockchain also has a digital signature for further protection. User-specific private and public keys are provided.

1.4.8.3 Mining

All transactions sent over the network are collected by miners, and only legitimate transaction that are forwarded to other nodes. Each miners add a number of gathered transactions to a freshly generated block.

1.4.8.4 Decentralization

Decentralization is a key characteristics of blockchain technology that are not dependent on or kept in the central area when something is decentralised. Instead, each Blockchain block has data that is stored there. The central authority does not distribute transactions to different nodes. The verifiable digital ledger is represented by each block. Cloud, IoT, edge computing, and big data are only a few study areas that use blockchain to do away with the idea of centralization and replace it with the idea of decentralisation.

1.4.8.5 Immutable

Something that cannot be altered is said to be immutable. Blocks cannot be changed, which is a crucial aspect of Blockchain. Immutability is attained using the proof of work idea. Mining produces tangible evidence of labour, and miners' work is to alter nonces. A nonce, which is a variable value used to create an exclusive Hash address for the block, is smaller than the desired hash value. Calculating proof of work has very little chance of happening. To obtain reliable proof of work, numerous experiments must be conducted. When the attacker simultaneously has control of more than 51% of the nodes, there is only one way to change the block.

1.5 IoT and Blockchain in Healthcare Systems

People from all walks of life, including attackers and retaliators, are drawn to a patient's medical history because it contains personal and sensitive information. Such information would be protected and disseminated with caution. A huge storage infrastructure is required in Medical IoT for real-time processing due to a large number of medical records. The bulk of medical IoT companies is currently storing and installing their application servers on the cloud. As previously said, data privacy and security are the major concerns when using the cloud to implement Medical IoT. We can't risk it since data could be destroyed or altered if cloud servers aren't trustworthy. Important data is shared between devices, and data leakage is unavoidable. Blockchain has several advantages has just gotten everyone's attention (thanks to the success of Bitcoin). A blockchain is a cloud-based Medical IoT security solution [33, 34]. The use of blockchain to deliver safe healthcare and data supervision has recently sparked interest [35]. A temporary-proven distributed ledger (Blockchain) can help protect the Medical IoT by documenting digital communication exchanges. Existing architecture that protects medical data using blockchain technology serves as a workaround for the safe transfer of patient health reports. A decentralised blockchain-based solution would deal with many of the issues related to the centralised cloud architecture. A blockchain-based data structure is a collection of nearly incorruptible cryptographically connected blocks that can be used to store sensitive patient information. The blockchain-based system is made up of connections between computers and people. The doctor is virtually present at a distance, keeping an eye on the patient's actions and giving them advise. The generated records are being examined by the doctor in the diagnostic centre as well. Electronic Medical Records (EMRs) are uploaded by the medical precisionist at the diagnostic centre and then saved to the patient's medical file. A specific clinic produces real-time statistical reports, which are then exchanged on a distributed ledger and reviewed by a medical expert. In order to keep an eye on the patient, the practitioner also employs wearable tracking devices. Wearable technology monitors changes in the patient's body and sends real-time information to the doctor. The physician then gave the patient counsel depending on the circumstances. The medical history of the patient is also visible to the patient's carers that any node in the patient network can read the patient's reports and treatment because they are stored on the distributed ledger. Healthcare workers use wearables to keep track of a patient's condition [36] and sensors are embedded in these gadgets, allowing them to monitor the patient at all times. Medical IoT can deliver crucial data to medical practitioners at any time.

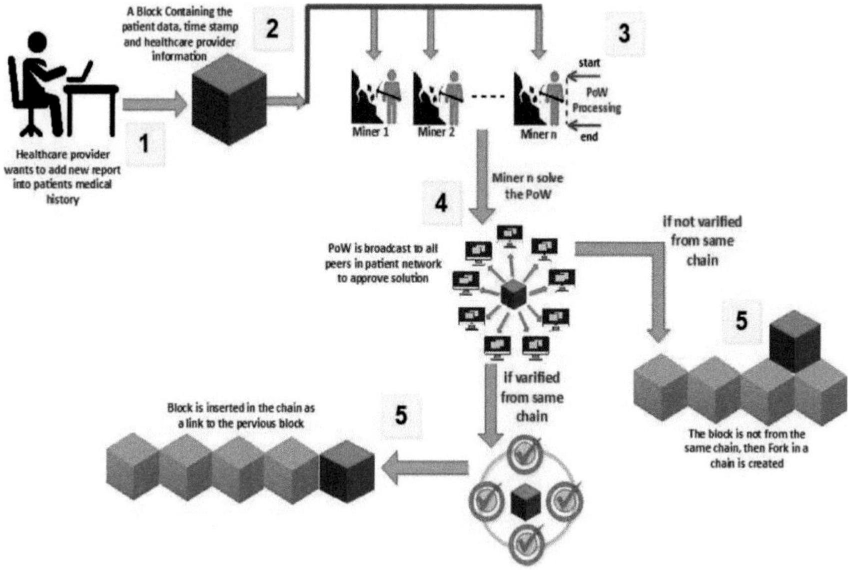

Fig. 3. Adding blocks.

A Blockchain-based, IPFS-based, and IoT-based secure Healthcare System

We outline the suggested system architecture and its various features in this section.

1.5.1 System Architecture

A Blockchain network connects the patient and medical team sides of the proposed system architecture, and both of them are able to communicate with the IPFS side. The system architecture contains three sides: patient side, medical team side, and IPFS side. The Blockchain holds a hash connection to the health data, which is encrypted and saved in the IPFS.

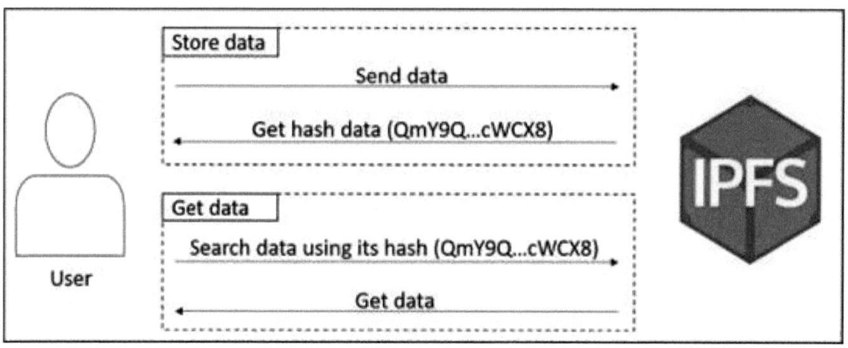

1.5.1.1 Patient Side

There are 2 types of devices in it:

A collection of Internet of Things medical devices and intelligent wearables with built-in sensors are available for every patient. These sensors can track various vital sign measurements as well as activity and sleep habits. As a result, the goal of these devices is to collect patient health data for transmission to the medical team via the patient's smartphone. Each of these devices is registered on the Blockchain by the owner (patient), and it can be recognised in the network by two identifying principles: its Source mac as well as the identity of the its own.

The medical personnel and IoT medical gadgets communicate with one other via smartphones. The node therefore provides access to the Blockchain for other Internet of Things devices. The smartphone won't store the entire Blockchain because of its limitations; instead, it will only have access to it via a Dapp. A Blockchain account is created for the patient by the Dapp when he creates his profile, and a pair of keys is established to serve as his special identification.

1.5.1.2 Health Care Team Side

It consists of doctors, hospitals, drug testing facilities, and public health agencies. Through a Blockchain network, these organisations are linked to patients and have access to their health information. Through their PCs or smartphones, doctors can access the Blockchain as light nodes. Hospitals function as complete nodes when it comes to maintaining a replica of the blockchain and participating in the consensus procedure.The Blockchain is available to the other entities, but they are not permitted to take part in the consensus procedure. For purposes of remote monitoring, analysis, and research, the patient data will be used.

1.5.1.3 IPFS Side

To store encrypted health data on this side, we employ IPFS, a distributed file system that is peer-to-peer, as an offchain database. For two reasons, we opted for IPFS instead of Blockchain to store our data. The system, in the first place, is capable of handling enormous volumes of data generated by several devices. This much data will need to be stored on dedicated full nodes, which will increase the size of the Blockchain and have an influence on its capacity. So, in order to protect data integrity and control access, the Blockchain is used entirely. Even though it is technically possible, it is not advisable to keep sensitive data on the Blockchain.

Since data are immortally kept on the Blockchain, all nodes will be able to decode and access all patient data if the encrypted system is ever compromised.

1.5.2 Proof of Authority

The Clique PoA [30] was used as a consensus method in this work because of the benefits it provides. On Ethereum, Clique has put the PoA into practise. The Geth Ethereum client is used to implement it, while Go was used for the construction process. In permissioned blockchains,Due to its ability to strengthen network security and consensus, PoA is among the most often utilised consensus algorithms. The adoption of such an algorithm has a lot of benefits. It works in part by accelerating consensus, which also accelerates data storage. This element is essential for applications in healthcare that require real-time processing. It is more energy-efficient, nevertheless, because it requires less processing power.

The PoA mechanism, in contrast to PoW, only permits a small number of specified validators, or authority, to take part in the consensus process by approving transaction and adding ever block onto the blockchain. In our case, hospitals serve as these validators. Since the PoA is based on a set of trusted authorities that are members of one or more well-known organisations, such the ministry of public health, it is typically utilised in the context of private and consortium Blockchains.

1.5.3 Ethereum

Ethereum is an open-source decentralised, shared computing platform built on the Blockchain. As a result of Vitalik Buterin drawing inspiration from the cryptocurrency Bitcoin, it was developed in 2014. Elliptic Curve Digital Signature Algorithm, which was created by FIPS 186-4, is also used by Ethereum (ECDSA). The discrete logarithm issue is the foundation of elliptic curve cryptography, which makes use of it to produce a set of keys. The following function, y2 mod p = (x3+7)mod p, determines the elliptic curve, or secp256k1 in Ethereum, where p 1 is a constant and the mod p designates that this curve is over an order p finite field. The variables x and y stand in for a point's secp256k1 curve coordinates.

In order to create a private key, one must first choose a random integer between 1 and 2256. The user's private key creates a digital signature for a transaction and attaches it when it is delivered to the Ethereum network. The public key is created by elliptic multiplying from of the private one using equation K = k*G, where K is the certificate authority produced from the secret keys k and G is the generating point. The legitimacy of the transaction can be verified by any network user, though.

The core goal of Ethereum is to provide Ethereum virtual machines (EVMs) that can run "smart contracts," a form of computer algorithm. A digital signature must match the sender's public key, or Ethereum address, in order to be regarded legitimate. Nick Szabo coined the phrase "Smart contract" for the first time in 1997. To formalise and safeguard interactions created through computer networks, smart contracts combine a number of protocols with user interfaces. Smart contracts are software components that must be deployed and updated on the Ethereum Blockchain network in order for the Ethereum Virtual Machine (EVM) to function. The output of these programmes, which transform a contractual logic into a set of rules and regulations, is sent to every network node.

In the event that the predetermined conditions are met, they are capable of self-execution. A smart contract has the ability to trade assets or values with other smart contracts as well as transactions and data. Additionally, Ethereum intends to develop and implement decentralised applications (Dapps). A Dapp's front-end is equivalent to the front-end of any other typical application. The two variants' back ends, however, are different from one another. The back end of a Dapp is built using blockchain smart contracts. Therefore, rather than being placed on a central server, the back-end is created using smart contracts and runs on the Blockchain network.

Use cases and practical scenarios

This section outlines the execution of our architecture after presenting a primary use case and a few real-world scenarios that are applied to a diabetes treatment case.

I. System functionalities

A patient won't be able to construct an identity after installing our Dapp until after hearing from his doctor. As a result, the only person with the authority to create and register a QR code for each of their patients is the doctor. We included the doctor's address in this QR code along with some other details about the blockchain network, such its chainID. After scanning the QR code, the patient may create an Ethereum address, which acts as both his private key and personal identification number, to start an account. The patient will be able to do a variety of activities using different Dapp interfaces after the Blockchain network has been established.

The smart contract not only validates the authentication but also our Dapp back-end and offers the several features depicted in the use case diagram in Fig. 4.

Implementation

The suggested approach is put into practise in this section. The system is built using an IPFS data storage system, a private ethereum blockchain network based on Clique PoA, and proxy re-encryption.

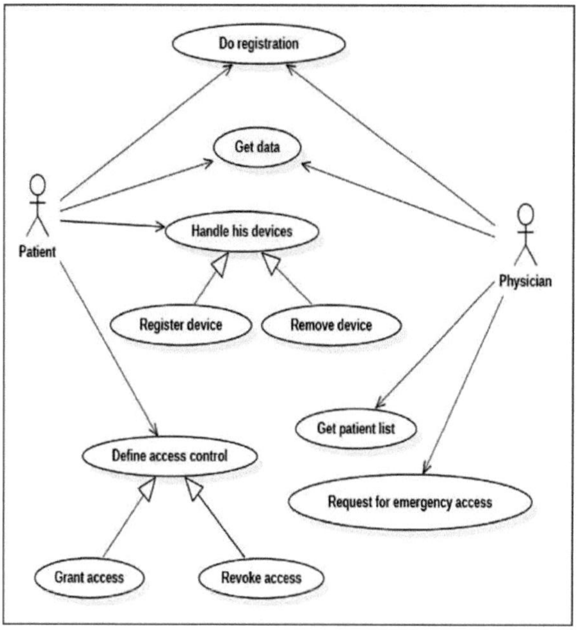

Fig. 4. Use case diagram.

II. Node types

We employed the Ethereum "Geth" clients architecture, as was previously indicated. Three different node types are present in the recommended solution:

- The "patient nodes" are smartphones. The patient Dapp, which links to our Ethereum Blockchain network via the Web3j API, allows these nodes to communicate with it. The network is connected using the JSON-RPC protocol.
- Physician nodes are another name for the computers and cellphones used by physicians. Patient nodes and the second kind are similar. The difference is that the connections to the Blockchain network are secured by the doctor's Dapp, not the patient's. In a manner similar to the previous, it hosts lightweight Ethereum clients.
- Hospitals networks are node that are connected to hospitals and public health organisations. They participate in the consensus process to approve transactions and add new blocks in their capacity as complete Ethereum nodes that store the whole Blockchain. We save the encrypted data on these nodes in a private IPFS network.

III. Data collection

Diabetes treatment requires a number of medical demands for a diabetic patient. Our activities, blood pressure, body weight, A1C and blood sugar levels, cholesterol and lipid levels, and blood pressure must all be monitored. The model for the glucose metre will be a Raspberry Pi. The patient's health evaluation and weight might be manually entered and uploaded using a smartphone. We'll use the sensors on a smartphone to monitor activities.

I. Data encryption/decryption

The NuCypher Umbral threshold proxy re-encryption technique [39] ensures both data encryption and decryption. The Python-based Umbral reference implementation, pyUmbral, was used in our research. In this study, we use Ethereum and Android to create mobile Dapps. Our Android project made use of the Chaquopy Python SDK to integrate pyUmbral.

Initially, the patient is given a set of keys (public and secret keys), and the doctor is given a separate set. The patient encrypts the data using his public key (). The user then creates the re-encryption key by using both his secret key and the doctor's public key (). The hospital (the complete node) receives the re-encryption key as soon as it is generated and uses it to re-encrypt the data on the patient's behalf. The medical centre attests to the doctor's authorization to access the information. The hospital re-encrypts the data using the re-encryption key without knowing the plaintext if the doctor has access rights.

Smart contract

A smart contract can be thought of as an automatic process that is started when a transaction is completed. All blockchain systems enable smart contracts, but they vary in terms of the languages that may be used to build them and the parameters that control their use.The most widely used programming languages for smart contracts include Java, Python, LLL, Golang, Solidity, and Serpent. The Haskell execution environment, Java Virtual Machine (JVM), and Ethereum Virtual Machine are just a few of the numerous possible execution environments (EVM).

Proof of burn: Validators using PoB "burn" coins by transferring them to a location where they will be permanently lost, as opposed to spending money on expensive hardware. Validators are given access to mine on the network based on a random selection procedure by sending the coins to an unreachable address. In this situation, the validators must make a long-term commitment in exchange for a fast loss. Miners may burn either as the native currency of the blockchain application or the currency of an alternative chain, like bitcoin, depending on how the PoB is implemented. Their chances of getting chosen to mine the next block increase as they spend more money. Although PoB is a fascinating alternative to PoW, it still waste power and resource needlessly.

Proof of Capacity: In the Proof of Capacity consensus, validators are supposed to use their hard drive space rather than burn coins or buy expensive equipment. If validators have a larger hard drive, they are more likely to be selected to mine the next block and get the block reward.

Proof of Elapsed Time: One of the most moral consensus algorithms is PoET, which choose which block to go on to based only on moral considerations. It is often used in Blockchain networks that have permissions. Any validator on the network has an equal opportunity to build their own blocks thanks to this strategy. For this, each node adds a proof of their wait to the block after waiting for a different length of time. In order for other users to evaluate the created blocks, they are broadcast to the network. When it comes to the proof part, the validator with the smallest timeout value prevails. Block uploaded to Blockchain by winning validator node.

Nodes cannot consistently win the election or provide the smallest timer value because of additional software protections. Leased Proof of Stake, Proof of Activity, Proof of Weight, Proof of Importance, and other techniques for reaching agreement are also available. Since consensus algorithms are required for blockchain networks to function as intended, it is vital to choose one carefully in line with the requirements of the business network.

Privacy preservation

The protection of privacy is the third topic covered in the literature. Data privacy and context privacy are the two main groups into which the privacy-preserving methods may be divided. Context privacy deals with spatial and temporal privacy whereas data privacy addresses issues such maintaining confidentiality while using data aggregation and querying methods.

Open issues

A survey of the literature identifies a few issues that require attention. In addition, it is clear from the arguments put forward in the papers and the research that has been done on them that a number of parameters regulate the integration approaches and data management activities in the two application domains. The constraints placed on an application area, the system's intended use, the number of IoT devices involved, the architecture employed, the application area itself, the sensitivity of the data, and the required throughput and latency are a few of these.

Within the parameters of those influencing factors, the reviewed research provide instantiations. While these publications offered many excellent answers, there are certain issues they did not cover and discrepancies amongst them about particular subjects, such as the following.

The several instantiations discussed here first set out to prove the prototype's superiority over competing ideas that were deemed important to the task at hand. When compiling this evaluation, the author discovered contradictions between the publications, such that what one magazine characterises as a strength is shown as a weakness in another publication. Consequently, it is vital to compare the effectiveness of these various instantiation types.

In the sections above, processing of data is categorised as an information management operation. Smart contracts enable more independent processing capacity in blockchain-based systems. It has been discovered, though, that different systems make use of various quantities of smart contracts. Future research can be based on determining how many smart contracts are appropriate for use on systems, as doing so might affect system performance.

Thirdly, the vast majority of publications fail to outline the best ways to obtain information. Due to its complexity, data retrieval in blockchain-based systems may occur either on-chain or off-chain. Image data are stored encrypted and on-chain in the healthcare sector. Data recovery from encrypted files, particularly the usage of searchable encrypted photo files, may be essential in certain situations. But this field of study has become stagnant. More research is thus required in this area.

The fourth issue is the disintegrative use of HIT, which is brought on by the lack of a trustworthy system for assigning unique IDs among other things. The qualities of blockchain is its ability to help's with problems like these, however it is rare to find literature that describes how blockchain may be used to connect various healthcare systems. Consequently, integrated HIT utilisation helps the domain; these problems might be seen as a permanent fix.

Fifth, there are other problems that require more study but have gotten less attention in publications. These concerns include how blockchain can be used to track disease transmission, patient monitoring, triage, and provider operational and financial performance.

1.6 Limitation and Solution

Blockchain can store information without unauthorized changes. However, there are still many obstacles to using blockchain in enterprises. Individual medical records are much larger than the vast bulk of public blockchain. Putting Private Medical Records (PHRs) into the blockchain is one of most difficult tasks. The blockchain will be incredibly large overall, making it challenging to focus on the vast amount of data that has to be processed. Blockchain was created to store small amounts of data in blocks (Bitcoin). Blockchain just maintains hash references to the data; the data itself would be kept in a separate off-chain capacity architecture to meet capacity constraints. Clinical data should be stored off-chain in accordance with the conventional data set design, however medical services data should be saved as changeless hashes on-chain to allow for the validation of access to off-chain clinical patient records.

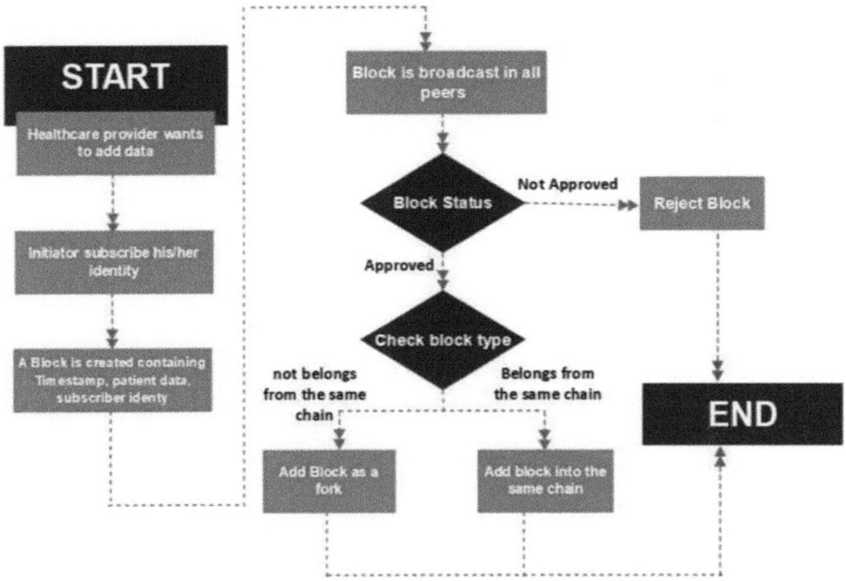

Fig. 5. Updating health care data.

2. Conclusion

Numerous business-related issues have changed as a result of the Internet of Things. One of the first industries to take advantage of this potential was the medical services sector, which did so by weaving a web of clinically related products. Due to the industry's quick development , diversity and security has become a major concern. Blockchain offers a safety and security assurance for Medical IoT. In order to cope with the constantly increasing number of IoT devices in medical services units, a decentralised system for doing so has been devised by connecting Blockchain with the IOMT (INTERNET OF MEDICAL THINGS), major amount of the security and protection risks are addressed by our suggested blockchain-based Medical IoT engineering because the information stored in the blockchain cannot be changed, a digital record of information based on checked agreements can be generated. From now on, blockchain ensures the confidentiality of patient clinical history is maintained at all times, and each hub in the Medical IoT organization is granted open admittance based on tempered evidence. In the future, we'll look into blockchain to see if there are any capacity difficulties.

References

[1] Agbo, C.C., Mahmoud, Q.H. and Eklund, J.M. 2019. Blockchain technology in healthcare: A systematic review, in healthcare. Multidisc. Digit. Publ. Inst., 7: 56.

[2] Ahram, T., Sargolzaei, A., Sargolzaei, S., Daniels, J. and Amaba, B. 2017. Blockchain Technology Innovations. IEEE Technology & Engineering Management Conference (TEMSCON).

[3] Esposito, C., Santis, A.D., Tortora, G., Chang, H., Kwang, K. and Choo, R. 2018. Blockchain: A Panacea for healthcare cloud-based data security and privacy? 2018 IEEE Cloud Computing.

[4] Hussien, H.M., Yasin, S., Udzir, N., Zaidan, A. and Bahaa, B. 2019. A systematic review for enabling of developing blockchain technology in healthcare application: Taxonomy, substantially analysis, motivations, challenges, recommendations, and future direction. J. Med. Syst., 43: 320.

[5] Kassab, M.H., DeFranco, J., Malas, T., Laplante, P., Destefanis, G. and Graciano Neto, V.V. 2019. Exploring research in blockchain for healthcare and a roadmap for the future. IEEE Trans. Emerg. Top. Comput., 1.

[6] Kizza, J.M. 2017. Internet of things (IoT): Growth, challenges, and security. In Guide to Computer Network Security, pp. 517– 531, Springer, Berlin, Germany.

[7] Forecast: The Internet of Things, Worldwide, 2013, Gartner, Stamford, CA, USA, Nov. 2013.

[8] Cisco: Enterprises Are Leading The Internet of Things Innovation, Aug, 2017.

[9] Barro-Torres, S.J., Fernández-Caramés, T.M., Pérez-Iglesias, H.J. and Escudero, C.J. 2012. Real-time personal protective equipment monitoring system. Comput. Commun., 36(1): 42–50.

[10] Fraga-Lamas, P., Fernández-Caramés, T.M., Blanco-Novoa, Ó. and Vilar-Montesinos, M.A. 2018. A review on industrial augmented reality systems for the industry 4.0 shipyard. IEEE Access, 6: 13358–13375.

[11] Forcast: Medtech and the Internet of Medical Things, July 2018, Deloitte Centre for Health Solutions.

[12] Rodrigues, J.J.P.C., Segundo, D.V.D.R., Junqueira, H.A., Sabino, M.H., Prince, R.M., Al-Muhtadi, J. and Albuquerque, D.V.D.R. 2018. Enabling Technologies for the Internet of Health Thing. IEEE Access, 6.

[13] Sun, W., Cai, Z., Li, Y., Lui, F., Fang, S. and Wang, G. 2018. Security and privacy in the medical Internet of Things: A review. Hindawi, Security and Communication Networks, Volume 2018, Article ID 5978636, 9 pages.

[14] WhitePaper: Cisco Visual Networking Index: Global Mobile Data Traffic Forecast Update, 2016–2021. San Jose, CA, USA, Mar. 2017.

[15] Jaimon T. Kelly, Katrina L. Campbell, Enying Gong and Paul Scuffham. 2020. The Internet of Things: JMIR publication published on 10.11.2020 in Vol 22, No 11.

[16] Nakamoto, S. 2008. Bitcoin: A peer-to-peer electronic cash system. Leslie Lamport, Robert Shostak, Marshall ACM Transactions on Programming Languages and Systems vol.4 issue 3| July 1982, pp. 382–401.Association for Computer Machinery New York, YN, US.

[17] Kathleen E. Wegrzyn Eugenia Wang. 2021. The Pros and Cons of Blockchain in Supply Chain, Foley & Lardner, 25 Aug 2021.

[18] Parvathavarthini, S. 2020. An improved crow search based intuitionistic fuzzy clustering algorithm for healthcare applications. Intelligent Automation and Soft Computing, 26(2): 253–260.

[19] Naresh, V.-S. 2020. Internet of things in healthcare: Architecture, applications, challenges, and solutions. Computer Systems Science and Engineering, 35(6): 411–421.

[20] Benjees. 2020. NASDAQ Partnership With Blockchain Firm R3 Is Great For Crypto. May 22,2020.Forbes.Available:https://www.forbes.com/sites/benjessel/2020/05/22/why-nasdaqs-partnership-with-r3-is-great-for-digital-asset-adoption/?sh=43c4e7d9630f.

[21] Yu, P., Xia, Z., Fei, J. and Kumar Jha, S. 2020. An application review of artificial intelligence in prevention and cure of COVID-19 pandemic. Computers, Materials & Continua, 65(1): 743–760.

[22] McGhin, T., Chop, K.-K.R., Liu, C.Z. and He, D. 2019. Blockchain in healthcare applications: Research challenges and opportunities. Journal of Network and Computer Applications, 135: 62–75.

[23] Ajaz, F. Mohd Naseem 1, Sparsh Sharma 1, Mohammad Shabaz 2, Gaurav Dhiman 3 COVID-19: Challenges and its technological solutions using IoT. Current Medical Imaging PMID: 33588738 DOI: 10.2174/1573405617666210215143503.

[24] Li, J., Cheng, J., Xiong, N., Zhan, L. and Zhang, Y. 2020. A distributed privacy preservation approach for big data in public health emergencies using smart contract and SGX. Computers, Materials & Continua, 65(1): 723–741.

[25] Miraz, M.H. and Ali, M. 2018. Applications of blockchain technology beyond cryptocurrency. Annals of Emerging Technologies in Computing (AETiC), 2(1).

[26] Halpin, H. and Piekarska, M. 2017. Introduction to Security and Privacy on the Blockchain. European Symposium on Security and Privacy Workshops (EuroS&PW). IEEE Computer Society.

[27] Malamas, V., Dassaklis, T., Hatzinikolaou, P., Burmester, M. and Katsikas, S. 2019. A forensics-by-design management framework for medical devices based on blockchain. *In*: 2019 IEEE World Congress on Services (SERVICES), vol. 2642–939X, pp. 35–40, July 2019.

[28] Griggs, K.N., Osipova, O., Kohlios, C.P., Baccarini, A.N., Howson, E.A. and Hayajneh, T. 2018. Healthcare blockchain system using smart contracts for secure automated remote patient monitoring. J. Med. Syst., 42: 1–7.

[29] Uddin, M.A., Stranieri, A., Gondal, I. and Balasubramanian, V. 2020. Blockchain leveraged decentralized IoT health framework. Internet Things 9: 100159. Available:https://doi.org/10.1016/j.iot.2020.100159.

[30] Dwivedi, A., Srivastava, G., Dhar, S. and Singh, R. 2019. A decentralized privacy-preserving healthcare blockchain for IoT. Sensors 19 vol 19,issue 2| Published: 15 January 2019, 326.

[31] Attia, O., Khoufi, I., Louis, A. and Adjih, C. 2019. An IoT-blockchain architecture based on hyper ledger framework for healthcare monitoring application. *In*: 2019 10th IFIP International Conference on New Technologies, Mobility and Security (NTMS), pp. 1–5, June 2019.

[32] Selim, M. and Elgazzar, K. 2019. BIoMT: Blockchain for the internet of medical things. *In*: 2019 IEEE International Black Sea Conference on Communications and Networking (BlackSeaCom), pp. 1–4, June 2019.

[33] Khatoon, A. 2020. A blockchain-based smart contract system for healthcare management. Electronics, 9: 94.

[34] Mathew, S., Gulia, S., Singh, V. and Dev, V. 2017. A review paper on cloud computing and its security concerns. Intelligent and Computing in Engineering, pp. 245–250 ACSIS, Vol. 10 ISSN 2300- 5963.

[35] Zhang, J., Xue, N. and Huang, X. 2016. A secure system for pervasive social network-based healthcare. IEEE Access, 4: 9239–9250.

[36] Kheer, S., Moniruzzaman, M., Yassine, A. and Benlamri, R. 2019. Blockchain technology in healthcare: A comprehensive review and directions for future research. Appl. Sci., 9(9): 1736 26 | April, 2019.

[37] Saha, A., Amin, R., Kunal, S., Vollala, S. and Dwivedi, S.K. 2019. Review on blockchain technology-based medical healthcare system with privacy issues. Secure. Priv., 2: e83. DOI:10.1002/spy2.83.

[38] Kshetri, N. 2017. Can Blockchain Strengthen the Internet of Things. IEEE Computer Society.

Chapter 4

Medical IoT Applications using Blockchain

David Al-Dabass,[1] *Naveena P*,[2,*] *Pradeepika T*,[2]
Nethra B[2] and *Padmashree A*[2]

Block chain is another innovation that is being utilized to foster novel arrangements in an assortment of fields, including medical services. In the medical care framework, a Block bind network is utilized to store and share patient information across emergency clinics, symptomatic research centers, drug stores, and clinicians. In the clinical calling, block chain applications can appropriately distinguish genuine and, surprisingly, dangerous blunders. Therefore, it can possibly build the presentation, security, and straightforwardness of clinical information partaking in the medical care framework. Clinical organizations can utilize this innovation to acquire knowledge and work on the examination of clinical records. There are fourteen significant Block chain utilizes in medical services. In clinical preliminaries, block chain assumes a basic part in identifying misleading; here, the innovation's true capacity is to further develop information proficiency for medical care. It can assist with mitigating worries about information control in medical services by empowering an exceptional

[1] School of computing and informatics, Nottingham, Trent University.
[2] Bannari Amman Institute of Technology, Sathyamangalam.
Emails: David.al-dabass@ntu.ac.uk; pradeepika.cb20@bitsathy.ac.in; nethra.cb20@bitsathy.ac.in; padmashreea@bitsathy.ac.in
* Corresponding author: naveena.cb20@bitsathy.ac.in

information stockpiling design with the most elevated level of safety. It gives information access assortment, interconnectedness, responsibility, and confirmation. Wellbeing records should be kept up with protected and classified for an assortment of reasons. Block chain empowers decentralized information assurance in medical services while likewise keeping away from interesting risks. IoT further develops correspondence among specialists and patients in medical services by permitting patients to be analyzed somewhat in crisis circumstances by means of body sensor organizations and wearable sensors. Involving IoT in medical care frameworks, then again, can be useful. Patients' protection might be disregarded, thus, it's critical to ponder security. Block chain is a famous report subject nowadays, and it very well might be utilized in a wide scope of IoT situations. Decentralization, Immutability, Security and Privacy, and Transparency are only a couple of the vital advantages of executing Block chain in medical services frameworks. The significant objective of this article was to work on the activity of medical services frameworks by consolidating arising and inventive processing advancements like the Internet of Things (IoT) and Block chain. At last, the challenges of consolidating IoT and Block chain into medical services frameworks are featured. We explored Block chain innovation and its huge advantages in medical care in this report. Diagrammatically, the numerous capacities, empowering agents, and brought together work process interaction of Block affix Technology to help worldwide medical care are examined. At long last, the review distinguishes and talks about various issues [1].

1. Introduction

The Healthcare locale is a chief pressure for all the making as well as cutting edge nations since this area is straightforwardly worried about the social government help and existence of people. Creative work in the health industry district could be a predictable cycle, as they assist with working on living by battling different clinical related problems and pollutions [1]. With the development and persistent movements in progression, the enhancement in the health industry locale should be clear without any problem. The Internet of Things (IoT) is a term used to portray an organization of associated gadgets. Everything being associated with the Internet is an idea. Vehicles, home apparatuses, and different things hardware, programming, sensors, actuators, and associations incorporated in merchandise [13]. This classification incorporates all that permits these things to associate, gather, and offer information. Kevin Ashton is eminent as the "Father of IoT," an abbreviation for Internet of Things reaches out to a wide scope of gadgets, including PCs, PCs, cells, and tablets [9]. Beforehand idiotic or non-web empowered actual things and ordinary items are currently accessible. Sensors, The most incessant advancements utilized in the cloud incorporate cloud, remote innovation, and security, Internet of things.

The fundamental IoT life cycle comprises of four stages: gathering information through sensors on gadgets, putting away the information in the cloud for examination, communicating the broke down information back to the gadget, and following up on the information. The Internet of Things is gainful in a huge number of ventures, making our lives simpler. The most predominant IoT applications are brilliant homes, savvy urban areas, agribusiness, shrewd retail, driverless vehicles, and medical services. Security is a fundamental part of any innovation, and it is particularly indispensable in IoT organizations [3]. Information classification and confirmation, access control inside the IoT organization, protection and trust among IoT gadgets are largely subjects of continuous review pointed toward improving IoT security.

Security and protection arrangements, as well as clients and things, are totally upheld. The IoT security issue emerges because of awful program plan, which prompts weaknesses, which are a vital wellspring of organization security issues. Block chain is a decentralized and open-source advanced record that records exchanges across various PCs so that no record might be modified reflectively without affecting succeeding squares. In the square chain, each, is confirmed and connected to the one preceding it, framing a persistent chain. All things considered, the record's name is Block chain. Block chain conveys an elevated degree of responsibility on the grounds that each exchange is recorded and confirmed freely. Nobody can vary the information written in the Block chain after this has been placed. Its motivation is to show that the information is current and unaltered. Information is put away on networks as opposed to a focal data set in Block chain, which further develops security while additionally presenting its weakness to hacking. Block chain gives a tremendous stage to creating and contending with customary enterprises for current and imaginative items [7].

The square chain innovation helps advertisers in monitoring clinical items. Using Block chain advances, the well being and drug businesses will actually want to take out fake meds while additionally taking into consideration the following of these items. It helps with the revelation of the creation's source. Whenever a clinical history is made, Block chain might save it, and this record can't be altered. This decentralized organization

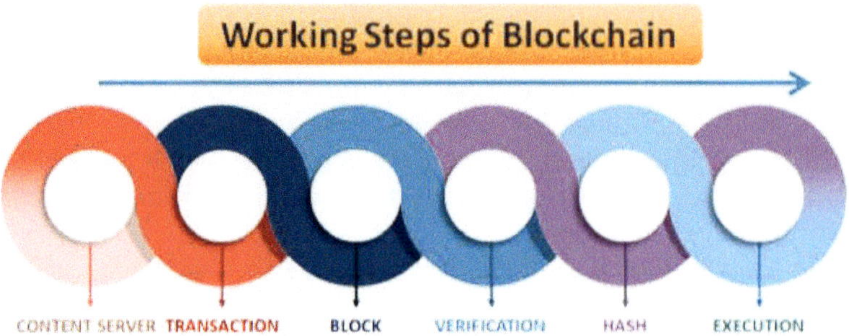

Fig. 1. Shows the basic working steps of blockchain.

interfaces all of the clinic's item equipment. Analysts can make gauges for treatments, medications, and solutions for a scope of sicknesses and issues utilizing the assets saved by these devices.

1.1 Need of Blockchain in Healthcare

Some of the biggest issues with block chain applications in healthcare are as follows:

- Security is critical at all stages of the network infrastructure.
- The identities of all participants are checked and authenticated.
- Consistent authorization patterns for access to electronic health records.

Despite the fact that DLT can be employed in a variety of healthcare settings, not all healthcare activity is transactional. However, because the data in public block chains is publicly accessible, they can't be used to store private information like identifying health information. Providers are expected to investigate privacy issues as a result of this transparency in order to ensure that protected health information is preserved (PHI).

Second, while block chain technology is vulnerable to some types of attacks, it also has built-in defenses for others. Because of its code, the block chain is susceptible to zero-day attacks, vulnerabilities, and social engineering. As a result, information security must be given considerable attention, particularly in the healthcare industry [2].

Because block chain data is immutable, it should not be used in healthcare on a whim. Large files, especially those with frequent changes, may be excluded. Personal information should be kept off the chain at all times.

Patient security is currently a need while thinking about handling any sort of information, as per DLT specialists, with new guidelines on the ascent, like the General Data Protection Regulation (GDPR), and guidelines that have been set up for over 10 years, like HIPAA.

The benefits of utilizing block chains over conventional strategies for medical care data set administration frameworks incorporate decentralized administration, unchangeable data sets, information provenance, detectable information, hearty information, and information accessibility to any approved client while keeping it out of the hands of unapproved clients through encryption that is subject to a patient's confidential key [5].

2. Blockchain Technology's Various Capabilities to Support Global Healthcare Culture

In clinical benefits, block chain has a wide extent of uses and limits. Record development helps clinical benefits experts in revealing genetic code by supporting the strong trade of patient clinical records, dealing with the medicine creation organization, and working with the safeguarded transmission of patient clinical records. Figure 2 depicts the various qualities and tremendous facilitators of Block chain thought in different clinical benefits fields and related disciplines. Medical services information assurance, different genomes the executives, electronic information the board, clinical records, interoperability, digitalized following an issue flare-up, etc., are a portion of the in fact determined and fabulous perspectives used to develop and practice Block chain innovation. Block chain innovation's absolutely computerized components, as well as its utilization in medical services related applications, are significant drivers in its reception.

From assembling to drug store retires, the Block chain unveils the whole solution process. IoT and Block chain can be utilized to follow traffic, cargo course, and speed. It empowers proficient procurement planning for centers, drug stores, and other clinical organizations that utilize a specific medication to keep away from disturbances and deficiencies. The utilization of Block chain-based computerized structures would help with forestalling unapproved adjustments to calculated information. It imparts trust and forestalls unapproved admittance to records, installments, and solutions by a wide scope of individuals keen on buying drugs. The strategy can productively work on patients' circumstances while holding assets at a sensible expense. It eliminates all impediments and limitations in staggered validation. Since Block chain takes into consideration a decentralized, upright record to be kept up with [20].

Besides, while Block chain is public, it is likewise private, concealing any singular's personality behind convoluted and secure calculations equipped for safeguarding the awareness of clinical information. In light of the innovation's decentralized construction, patients, specialists, and medical services suppliers may all have a similar data quickly and securely. Block chain innovation works with patient-drove interoperability by permitting clients to make their clinical information open and follow access necessities. This gives patients more command over their own data while likewise helping security and classification. It's trying to monitor and apply quality control and authorization. Block chain applications could take care of any of these mechanical issues in all cases. Administrative offices will be helped by the utilization of square chain features in following lawful medications and recognizing them from fakes [15].

This guarantees that all gatherings associated with computerized exchanges, including the patient's data, have been supported. Patients who change suppliers can basically refresh a solitary agree structure to have each of their clinical records moved. Block chain has advanced into the medical care industry, with a rising reception rate. In the beginning phases, individuals in the wellbeing environment likewise acknowledge the innovation well [19]. The general objective of Block chain in the medical services industry before very long will be to resolve gives that are presently influencing the current construction. It permits specialists, patients, and drug specialists to get moment admittance to all suitable data. At the entire hours of the constantly, clinical firms are contemplating, attempting, and finding Block chain innovation in the clinical field for wellbeing records. It has demonstrated to be a significant instrument in the domain of medical care.

Notwithstanding complex innovation, for example, AI and man-made consciousness, the clinical business is essentially dependent on Block chain. Block chain's change of the medical services industry has a few genuine applications. The application utilizes Block affix checking advancements to tweak the restorative store network. The square chain innovation's true capacity takes into consideration a convoluted information stockpiling framework that monitors an individual's entire clinical history, including analyze, test results, past medicines, and, surprisingly, astute sensor measures. A specialist can quickly accumulate all of the data expected to make exact judgments and suggestions utilizing this strategy. Since the information in a solitary Block chain framework is saved, it is protected against misfortune and change. The square chain innovation can be utilized to evade an association's inner organizations [17].

A scrambled Block chain data set can protect an immense association with numerous unmistakable partners and levels of power from outer dangers and assaults. In the event that a medical care association effectively executes a Block chain organization, salvage assaults and different challenges, for example, PC debasement or equipment disappointment will be disposed of.

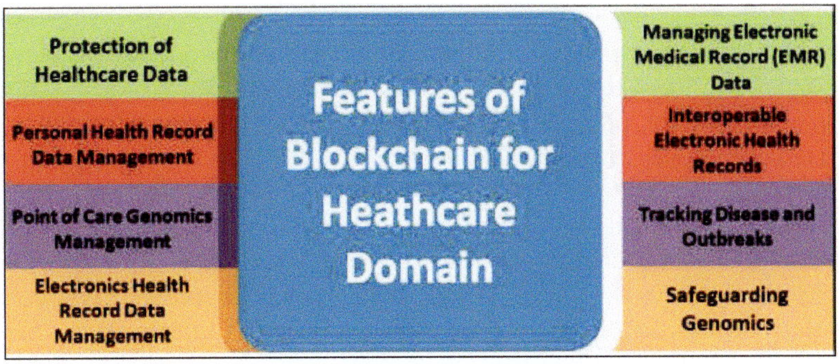

Fig. 2. Limits of block chain innovation for medical care area.

3. A Few Blockchain Applications in the Health Sector

Despite some progress in published research on the application of block chain in the health care industry, the current state of the art is still immature. Although most research focuses on new block chain frameworks, ideas, or models, technical aspects regarding the technology are also important. The block chain components that are used are rarely disclosed. It is uncommon to come upon block chain technology. On a national basis, people are employed in health care. Some countries, such as Estonia and Malta, have adopted this policy and demonstrated how block chain technologies could aid data security and patient consent management. Identity management is one of the most promising applications of block chain in the health-care sector like medical supply chain management, dynamic patient consent, and management of medical supplies as well as medications [12].

3.1 Blockchain and Internet of Things (IoT) in Healthcare

Consistently, the quantity of patients in the United States develops, making complete clinical consideration harder to offer. Because of the utilization of IoT and wearable gadgets as of late, the nature of clinical consideration has moved along. The most widely recognized technique for managing medical services issues is distant patient observing. Wearable devices that accumulate and move information to medical clinics, as well as Internet of Things (IoT) gadgets, have a significant influence in far off quiet checking. The essential objective of these gadgets is to offer indispensable data to medical care suppliers, for example, an individual's breathing examples, blood glucose level, and circulatory strain [11].

In medical services, there are four distinct sorts of information assortment gear: Medical Embedded Gadgets are gadgets that are carefully embedded into an individual's body. Wearable Health Monitoring Gadgets: These are medicinally recommended gadgets that are worn on the body. Clinical Wearable Devices: Doctors recommend these gadgets. RPM's significant objective is to safeguard information that is being focused on by programmers. Block tie innovation is used to safeguard the information. By using the decentralization guideline, block bind assists with getting information from an assortment of dangers. The information is likewise checked utilizing block chain savvy contracts [9].

Medical services are a sort of IoT framework with its own arrangement of safety necessities, including interoperability and information sharing. The most common way of trading information between various sources is known as interoperability. The incorporated idea has the downside of being not able to accomplish interoperability [22]. The Internet of Things is predicated on uncertain brought together information stockpiling on the cloud. The security gives that medical services applications defy can be settled by joining block chain and IoT. Many Block chain preliminaries have proactively been led in the medical services industry.

4. The Future of Blockchain in Healthcare

In the domain of clinical innovation, block chain is quite possibly of the most conspicuous trendy expression. There's a sensible clarification for this. Essentially said, block anchor can possibly change the medical services area. Once completely embraced, patients will be set at the focal point, everything being equal, which will be totally redesignd with expanded security, protection, and availability. In any case, how is all of this made conceivable by block chain? How is the medical care industry using this state of the art innovation to its maximum capacity [18]?

4.1 What Impact Does Blockchain Have on the Medical Industry?

Because block chain is open and secure, it may be utilized in a range of medical applications, resulting in considerable cost reductions and new ways for people to access treatment.

When combined with the compounding nature of data and creativity, future proofing technology can be leveraged to create an era of growth and innovation [17].

The framework for a block chain revolution is already being laid by pioneering enterprises. Here are some of the ways they're making an impact in medicine.

4.2 *A Future Fueled by Innovation*

The foundation of expansion and adaptability that block chain technology enables is one of the most fascinating elements of the technology. Though it will encounter the same restrictions as today's technology in its initial application, block chain's open nature will stimulate and support industry-wide innovation for many years to come.

According to BIS Research, by 2025, the rapid implementation and integration of block chain in healthcare could save more than $100 billion in costs linked to IT, operations, support functions, staff, and health data breaches.

While many innovative and intriguing block chain solutions have emerged from leading organizations around the world, this is just the beginning. Join the revolution today and help push the boundaries of medical technology while setting the groundwork for tomorrow [23].

5. Blockchain's Unifed Work-Flow Process Technology Realization in Healthcare Amenities

Consistently, the quantity of patients in the United States develops, making complete clinical consideration harder to offer. Because of the utilization of IoT and wearable gadgets as of late, the nature of clinical consideration has moved along. The most widely recognized technique for managing medical services issues is distant patient observing. Wearable devices that accumulate and move information to medical clinics, as well as Internet of Things (IoT) gadgets, have a significant influence in far off quiet checking. The essential objective of these gadgets is to offer indispensable data to medical care suppliers, for example, an individual's breathing examples, blood glucose level, and circulatory strain [21].

In medical services, there are four distinct sorts of information assortment gear: Medical Embedded Gadgets are gadgets that are carefully embedded into an individual's body [10]. Wearable Health Monitoring Gadgets: These are medicinally recommended gadgets that are worn on the body. Clinical Wearable Devices: Doctors recommend these gadgets. RPM's significant objective is to safeguard information that is being focused on by programmers. Block tie innovation is used to safeguard the information. By using the decentralization guideline, block bind assists with getting information from an assortment of dangers. The information is likewise checked utilizing block chain savvy contracts.

Medical services are a sort of IoT framework with its own arrangement of safety necessities, including interoperability and information sharing. The most common way of trading information between various sources

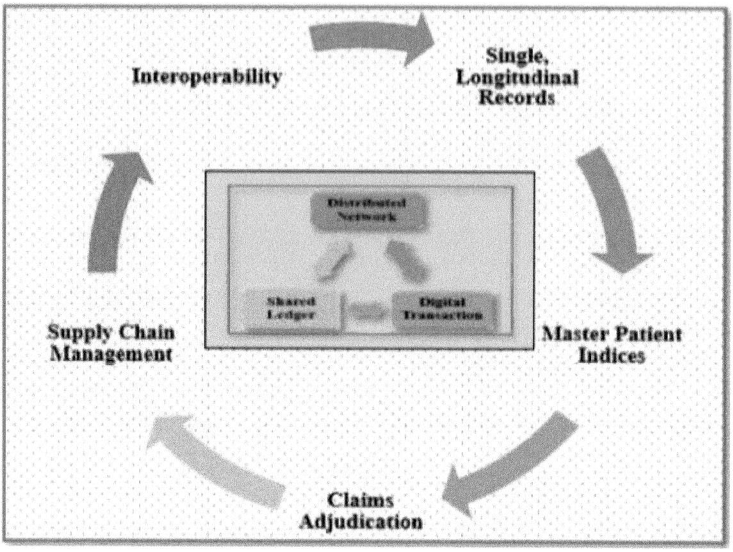

Fig. 3. Coordinated work process cycle of block chain innovation for medical services culture.

is known as interoperability. The incorporated idea has the downside of being not able to accomplish interoperability. The Internet of Things is predicated on uncertain brought together information stockpiling on the cloud. The security gives that medical services applications defy can be settled by joining block chain and IoT. Many Block chain preliminaries have proactively been led in the medical services industry.

6. Limitations and Future Scope

Block chain innovation is being utilized in the medical services industry, which has its own arrangement of issues. A critical obstacle to the arrangement of this advanced innovation in clinical offices is an absence of ability. Block chain applications are as yet in their beginning phases, and more mechanical innovative work is required [14]. It does, in any case, relate to clinical gatherings' and controllers' liabilities. The medical care framework has for some time been needing improvement. Block chain's use in medical services is very liable to grow from here on out. Because of this specialized improvement, its applications in medical services will improve, as it helps with the clarification of therapy results and progress. For approving exchanges and information moves, block chain innovation is critical. In the next few days, exchanges will be validated and enrolled utilizing Block chain advancements, with network individuals' consent.

Block chain will guarantee mathematical security to patients by means of public and private key encryption as the underpinning of another period of wellbeing data trade. This innovation is projected to help patient records, encroachment avoidance, interoperability improvement, process justification, medication and solution control, and clinical and production network checking. Later on, block chain in medical care is projected to perform staggeringly well [11].

7. Medicalcare Datasharing through the Gem Health Network

The clinical consideration industry handles many records and documents that are confidential and fall under extreme guidelines, for instance, Health Insurance Portability and Accountability Act of 1996 (HIPAA). Such records are generally taken care of in a concentrated informational collection, which could achieve issues, for instance, security and interoperability inspected in the previous section. The difficulty of sharing private records is an issue can happen when a patient necessities aggregated therapy at another crisis facility, thruway or abroad (for instance right when a 'clinical the movement business' action ends up being terrible in an abroad clinical center and the patient requires desperate clinical treatment). Each time the patient sees another capable, there is an interference due to the need of making new records, changing correspondence shows between trained professionals, reviving of various clinical prosperity records, and conflicting IT interfaces that can provoke monotonous and resource genuine approval and information taking care of for all social occasions included.

A logical solution for this issue has been made by Gem Health Network [25] that used the Ethereum Block chain Technology to make a typical association establishment. The Gem Health Network licenses clinical benefits specialists to generally get to comparative clinical benefits information disposing of the constraints of concentrated storing. The clinical records and information are huge, direct, and endorsed clients have continuous induction to the latest treatment information. This can help with restricting the bet of clinical thoughtlessness on account of old information and thwart clinical issues that could rise out of that lie.

Estonia has purportedly collaborated with Guard time, which is a clinical benefits stage that uses block chain development. This licenses Estonian occupants, clinical consideration providers, and medical care associations to recuperate all clinical treatments acted in Estonia by using the advancement. This prescribed the utility of block chain to be utilized in the manner discussed above [25].

The Gem Health Network tries to determine the issue of splitting patient records in a steady environment between the two patients and specialists. There are, regardless, different moves that ought to be settled, similar to character the leaders for the patient and expecting there is any sort of key organization to get the records from adjusting and abuse. Besides, accepting a public key is lost or spilled, how should the information be recovered and the compromised key is invigorated [19]?

8. Application of IoT and Blockchain Technologies in Healthcare

The quantity of patients in the United States is developing consistently, and giving complete clinical treatment has become more testing accordingly. With the utilization of IoT and wearable gadgets, the nature of clinical consideration has expanded as of late. The essential strategy for tending to medical services issues is far off quiet checking. Wearable devices that assemble and communicate information to clinics, as well as Internet of Things (IoT) gadgets, assume a basic part in distant patient observing. The significant objectives of these gadgets are to offer essential data to wellbeing specialists, for example, an individual's breathing examples, blood glucose level, and circulatory strain [7].

Information gathering gear in medical services can be partitioned into four classifications: Medical Embedded Devices: these gadgets are embedded into the human body, Medical Wearable Devices: these gadgets are endorsed by specialists, and Wearable Health Monitoring Devices: these gadgets are worn on the body. RPM's significant objective is to safeguard information that is being designated by programmers. Blockchain innovation is used to protect the information. By using the standard of decentralization, blockchain assists with shielding information from an assortment of dangers. Brilliant agreements on the blockchain additionally validate the information.

Medical services is a sort of IoT framework that has its own arrangement of necessities, for example, interoperability and information move, for getting patient information. The most common way of imparting information to various sources is alluded to as interoperability. The impediment to achieve interoperability is important for the incorporated idea [10]. The Internet of Things is predicated on centralization, with information being put away in the cloud, which isn't protected. The security worries that medical care applications experience can be overwhelmed by consolidating blockchain and IoT. Many examinations in Blockchain have proactively been led in the field of medical care.

9. Challenges of using the Blockchain in Healthcare

Despite the fact that block chain is a multidisciplinary thought with hindrances and limits, it tends to be applied to an assortment of fields. Analysts in this field are attempting to survive or moderate the adverse consequences of these components. We've accumulated a synopsis of a portion of the issues (i.e., mechanical difficulties) that the organization is confronting.

When block chain innovation is utilized in medical care:

- Throughput: As the quantity of exchanges and hubs in the organization develops, so does the throughput. More checks should be performed on the organization, maybe causing an organization blockage. While working with medical services frameworks, high throughput is an issue since there isn't sufficient opportunity to do everything. Given fast access, this could endanger a finding that could save somebody's life [13].
- Inactivity: Validating a square requires approximately 10 minutes; this can be hurtful to framework security administrations since effective attacks could happen during that time [15].

  Medical care frameworks are dynamic and ought to be gotten to consistently, as any postponement can risk a test's examination.
- Security: If a substance deals with 51% of the organization's handling power, this can be endangered. This is a significant issue that should be addressed since a hurt medical care framework can prompt medical services associations losing their validity [16].
- Asset utilization: Because the mining system consumes a ton of energy, utilizing this strategy could bring about a huge loss of assets. In a medical services setting, in light of the fact that various gadgets are expected to screen the patients, the energy costs are incredibly high; by the by, the use of block chain may likewise bring about critical processing and energy costs. Dealing with these expenses is quite difficult for organizations [18].
- Ease of use: Because these frameworks are so muddled to deal with, ease of use is additionally a worry. Moreover, shoppers will require an API (Application Programming Interface) with easy to use highlights. Since medical care suppliers don't have similar degree of specialized skill as IT subject matter experts, the frameworks ought to be basic and clear.
- Centralization: Despite the way that block chain is a decentralized plan, certain methodologies are unified. Therefore, the excavators

will generally be concentrated, which influences network unwavering quality. Since this focal hub is uncertain and can be hacked, the information it contains is in danger. Malignant attacks can gain admittance to the framework.

- Protection: It's generally accepted that the Bitcoin structure permits block chain to ensure security. Its hubs' security is safeguarded. The discoveries of, then again, discredit this reason.

Besides, answers for give this ability to block chain-based frameworks are required.

Because of security rules and guidelines, block chain-based frameworks should comply with them [23].

The European Union's General Data Protection Regulation (GDPR).

10. Healthcare Blockchain Privacy and Security

Since actual records have been digitized, and sensors and other innovative hardware are utilized to create wellbeing information, there has been a development in wellbeing information. In the medical care industry, information is created from an assortment of sources, including emergency clinic records, radiographic pictures, and checking gadgets.

There are different data sets with patient data, some of which might be delicate because of the way that it shows restraint data. Be that as it may, one of the techniques to guarantee more grounded security in these settings is to embrace block chain innovation, which has various benefits, including unchanging nature and information detectability apparatuses. The following sections go through a portion of the parts of block chain that relate to medical care security. Above all else, block chain-based medical care frameworks can further develop cryptography-based frameworks. Lays forward a methodology for finding the ideal cryptographic models for savvy contracts in his work. In any case, security worries with block chain stay unsettled. While sending information, there is a need to investigate a few security issues [8].

- Character security: keeping the client's very own data hidden and not connecting it to the exchange.
- Exchange security: ensuring that non-approved clients don't approach exchange content.

Since the information assembled in these records contains individual data, the usage of patient medical care records raises concerns concerning patient protection. An individual enrolment number, the quantity of the Visa that the patient purposes to make an instalment in the private organization, and other data might be remembered for these records. Looks analyses various procedures in the writing that can assist with these issues, for example, k-secrecy and zero-information evidence.

The creators present a conventional worldview for protecting security, which includes a procedure for demonstrating data's legitimacy without uncovering it to different hubs.

Coming up next are a portion of the security procedures to know about: (i) confined in execution conditions (TEE), (ii) homomorphic cryptography, and (iii) zk-snarks are generally instances of confided in execution conditions [9].

At last, in medical care conditions where records are shared, methods, for example, zero-information confirmation and quality based encryption can be utilized straight by alluding to the wellbeing record document or information put away in a data set to address security concerns.

It ought to be noticed that information protection is basic in the medical services setting, and block chain innovation can assist with guaranteeing this climate's security. Moreover, there is a current development in medical services frameworks to utilize encryption ways to deal with further develop security, which includes a block chain in view of differential protection. Differential security shields information from linkage assaults, in which related information from two data sets is connected, or from construing and finding patient delicate data [14].

11. Blockchain and its Features

Blockchain innovation has a large number of elements that can be used in the medical care industry. These highlights are innate to the framework and can be applied to an expansive scope of frameworks and enterprises. The highlights to be talked about explicitly in this segment are security, confirmation, and decentralized stockpiling [19].

11.1 Decentralized Capacity

Decentralized capacity is a significant component of the blockchain innovation and the reason for the improved security and verification of the data put away inside the framework [24]. Decentralized capacity is the most common way of separating the capacity of records from one significant server to various servers through blockchain's record, and can work with quicker admittance to clinical information, framework interoperability, patient organization, further developed information quality, and amount for clinical examination. Blockchain innovation can, for instance, be used by IoT and cloud suppliers to share information, both safely and secretly, in a decentralized way.

Information security, trustworthiness, and unchanging nature are centre elements of blockchain frameworks. Blockchain using its private secure, decentralized record can offer security by putting away data among numerous PCs rather than in a solitary source. Such a disseminated stockpiling strategy permits an exchange to be dispersed through the whole blockchain network making numerous excess information sources to check the credibility of the first exchange. By having this overt repetitiveness, a malevolent entertainer can't make any change without changing the data on every one of the frameworks inside the organization; consequently, considering permanence, confirmation and security [16].

Information security, uprightness, and protection are developing worries as they apply to brilliant gadgets, public cloud frameworks, and the overall IoT foundation. Blockchain can be, in principle, used to get IoT gadgets. For instance, a blockchain framework permits the cloud administrations to serve this edge has (for example brilliant or IoT gadgets) as near the gadgets as conceivable with a decentralized nature permitting added security and control. Blockchain additionally permits secure information sharing by giving information provenance, examining, and control for cloud-based server that store divided clinical information between enormous information substances. This can be achieved through savvy agreements and access control executions to help safely secure cloud-based foundations. Blockchain likewise has applications in innovation that permits client solicitation and waiter answer client-waiter frameworks by giving an auditable framework to add extra security [11].

11.2 Authentication

Blockchain through its decentralized framework, additionally guarantees the validation of records or other confidential data that is put away in blocks along the blockchain. Confirmation is achieved by requiring a particular confidential key that is attached to a public key to start the creation, modification, or review of data put away in the blockchain. These keys are put away in an assortment of programming called a Bitcoin wallet and is related with a Bitcoin address. These product applications are by and large applied towards Bitcoin and digital currency, however can be adjusted for other verification processes by utilizing a similar cryptographic plan. This validation is being explored in its capacity to verify character and personality documentation going from government archives to private medical services records [19].

11.3 Applications of Blockchain

A blockchain is undoubtedly a series of blocks obtained by cryptographic methods. The durability of this project is one of the overwhelmingly captivating aspects. Since the information above cannot be modified, it is possible to create an accurate record of the information that is based on understanding and is both clear and concise. Due to these factors, blockchain is particularly well suited to projects where information accuracy is of the utmost importance. Prochain, a blockchain-based system that enables chains to connect to cloud-based information objects, is a key example of this constancy. One of the ways that blockchain operations change is by understanding frameworks. For instance, Bitcoin uses a Proof-Of-Work algorithm called Hash Cash and has an intentionally high starting assessment designed to thwart organisation attack rejection [24]. By using this proof-of-work estimate as a vote in favour of the blockchain's settlement, all Bitcoin tractors support the blockchain. Similar to Bitcoin, Ethereum uses a Proof-of-Work algorithm called Ethash that takes into account the Dagger-Hashimoto calculation. Every centre point on the blockchain executes a piece of code known as a wise arrangement. They are self-executing contracts where the strategy is kept up to date across all users of the blockchain. They outline the advantages, obligations, and restrictions directly connected to an understanding of how a typical arrangement operates. As a result of their resemblance to the structures and regulations found in regular paper, they can be utilised, for example, to illustrate the clinical advantages of a person. Another type of blockchain trust model that is based on consortium trust is beginning to take shape. A structure dubbed Coco that takes into account the formation of blockchain freethinker consortiums was designed by Microsoft. These consortium agreements are based on a coalition of trustworthy parties. Various clinical facilities, including NHS Trusts in the UK, as well as producers of medicinal equipment and outcasts may fall within this category. By making superb preparations with just these reliable allies and equipment, understanding can be developed without the use of diggers. With a Coco-upgraded blockchain model ready to handle 1600 trades per second, this has happened in a fundamentally more substantial execution, bringing blockchain much closer to the enormous portion processors. 18 Coco is also a set of allowed execution circumstances, allowing the usage of, among example, Windows Virtual Secure Mode, Arm Trust Zone, and Intel Software Guard Extensions [22].

11.4 Clinical Trials

Blockchain has the potential to improve the accountability, transparency, and openness of clinical experts and trained professionals in the management of basic subject consent and clinical primers. Regulators can ensure that the foundation complies with the nation's prescribed consent regulations without the need for a really exceptional stretch screen clinical starter rule by maintaining an ongoing log of patient consent. This is especially significant because one of the most frequent forms of therapeutic deception has been the use of made-informed consent arrangements. This combines falsifying patient consent and altering medical records, demonstrating the need for some level of fundamental subject approval to stop this putting in place a clever agreement system that prevents physicians from using patient data until a key has been delivered at the end of an auditable splendid [20].

11.5 Data Sharing

The ideal entrance point for redesigns in clinical benefits may be information sharing locations, which also provide one of the biggest security issues. Powles and Hodson address the need for transparency regarding the distribution of patient data to third parties via a context-focused analysis of the DeepMind collaborative effort with the Royal Free London NHS Foundation Trust. Despite the positive impact Google's item suite has on persistent discovery and treatment, the lack of patient consent prior to logical testing is considered to be one of the biggest problems. IBM and the American Rest Apnea Association are focusing on rest apnea in a significant number of Americans at home utilising the IBM& Watson supercomputer at the far end of the reach and the patients' free, prior, and informed permission to manage significant clinical challenges. It is essential to have a national standard for interoperability in IT enterprises for clinical benefits. This was emphasised in a white paper for the NHS in the United Kingdom that Wachter and Hafter produced in collaboration with the US. Given that many trusts use various systems developed by various vendors to access these records, the clinical consideration structure highlighted the significance of interoperability in granting authorization to patient Electronic Health Records (EHRs) across various clinical facilities. Additionally, as evidenced in the U.S., this causes problems for clinical specialists and trained professionals. According to reports, social consideration and local wealth have persisted in the U.S. as well as the UK. A persistent problem in the clinical consideration area is clinical device asset following. According to a report by Harland Simon25 on an effort to legalise RFID marking in NHS Cambridgeshire, 15% of a clinical facility's assets are lost consistently, leading to a high expense for replacing items the crisis centre now owns. Additionally, a study conducted by GE Medical consideration discovered that orderlies spend an average of 21 minutes per shift looking for misplaced beds and devices. The study also concluded that any device costing less than $5,000 should be replaced if it cannot be found due to the high cost of doing so. With regard to clinical equipment according to radiofrequency conspicuous verification (RFID) regulations, NHS Forth. According to a further audit published by Harland Simon27, Valley in Scotland had the option of saving over £400,000 in cost repugnance by refusing to accept large therapeutic devices that would have been lost to the system anyhow. Drug monitoring completely replaces clinical device monitoring because fake medications are the main problem. According to a WHO study, up to 10% of the medicine supply in the United States may be fake. The Food and Drug Administration (FDA) in the US actually embraced the use of RFID to follow drugs from the stock organization to the patient. This enables the entire power chain to be examined, confirming that the facilities have purchased the medications from a reliable source [12].

11.6 Patient Records

Blockchain has the intrinsic power to challenge conventional wisdom and place responsibility for information. One particularly alluring step in this direction is MedRec, 30 which provides patients and professionals with a very reliable log of clinical benefits. It accepts an alternative method of aiding diggers by allowing the use of clinical benefit information that has been anonymised as a concession for maintaining the association. MedRec plans Patient-Provider Connections (PPRs) using smart contracts, and the understanding displays a brief summary of references relating to the interaction between focus focuses on the Blockchain. PPRs are also given patient responsibility, giving them the ability to understand, reject, or alter their relationships with clinical consideration suppliers like offices, financiers, and key interests. Blockchain provides a potential for clinical benefits frameworks to be interoperable. Blockchain presents an opportunity for clinical advantage framework interoperability by providing a decentralised record of perceived reality in clinical records that is accessible to all clinical advantage suppliers. This essentially assumes that, despite the UIs' potential for intrigue, their primary information will be ambiguous across all vendors. A test that is currently in place connects to the continuous success records across vendors, which include significant portions of the same data using different, potentially unrelated identities. Due to the duplication created by the blockchain and the resulting show contamination, it would be reasonable to assume that deduplication would maintain a sensible performant framework using new, anonymized identifiers to see patients across all relationships [18]. This is a business challenge without assistance from any other individual of taking on a blockchain thriving record, it is fundamental to see that flourishing records wouldn't begin from zero as they would need to uproot the continuous construction which makes difficulties [24].

11.7 Drug Tracking

Another entry is drug tracking on the blockchain, which exploits the stability of the ledger to support tracking and the chain of care from manufacturer to patient. In order to prevent the tampering and theft of prescriptions, Chronicled, a development new company, is enabling their product, Discover, which creates a chain of custody model that shows where the medication was delivered, where it has been since, and precisely when it was directed to patients. This permits providers of therapeutic advantages to comply with current clinical thinking regulations about the security of the tranquillizer supply, again with a supplement on interoperability between providers of clinical advantages. The Counterfeit Medicines Project was really launched by Hyperledger, the Open-Source Blockchain Working Group, focusing on the problem of bogus medications. The early stages of bogus prescriptions can be tracked using blockchain technology and eliminated from the hold chain. The advantage of blockchain in drug regulation over conventional methods is the decentralisation of trust and authority represented in the standards underlying the event, where focal specialists can be bought off or made to appear more knowledgeable than they actually are. In this way, Pedigree, the industry leader in drug detection, which now uses RFID and a common data base, is working toward their own blockchain strategy. Fake prescriptions might completely disappear from sharing reserve chains if medications could be modified and tracked using blockchain's natural nemesis of shifting cutoff points at the sign of gathering.

11.8 Device Tracking

Another opportunity for the blockchain to challenge healthcare thinking from the manufacturer to decommissioning is clinical device following. The NHS East Kent Hospital discovered that, according to a credible audit by Harland Simon35, in which they used strong RFID trackers on their high waste equipment, they discovered 98 mix syphons they didn't think they actually ensured over three locations. They saved $147,000 by using this single setting focused evaluation at a cost of $1,500 per person. The use of the blockchain in conjunction with this improvement provides the opportunity for a very robust record that demonstrates the location of the device, where it has been throughout its lifecycle, as well as which manufacturer, component, and continuous number are connected to the device, supporting administrative consistency. One of the regular key advantages for blockchain in the clinical advantages district was this cutoff, according to a white paper by Deloitte. In fact, a study by IBM36 found that 60% of government accessories in clinical advantages believe that resource allocation and clinical device split are the best areas for reducing local impact. A blockchain technique has two or three benefits over a traditional district-based approach of noting things. The stability and thoughtfully organised properties of the Blockchain are the most obvious of these. This prevents a malicious client from altering a device's district history or removing it from record. This is a particularly important component given that clinical device theft and shrinkage have escalated into major problems in both the US and the UK. Along with preventing general theft, its never-ending nature also prevents lost and replaced gadgets, which adds significantly to the cost of both the idea and the actual goods. Since all that would be needed to enter the device's consistent location would be to tap it with a PDA or scanner, this design shouldn't significantly increase the time required of a clinical manager, overseer, or sponsorship specialist. Although the Internet of Things (IoT) is rapidly adopting blockchain technology, yeah proposes a method for devices to connect with one another using an Ethereum blockchain and an RSA public key architecture. Thus, a device maintains its associated private key on the device and its associated public key on the blockchain. The NHS East Kent Hospital discovered that, according to a credible audit by Harland Simon35, in which they used strong RFID trackers on their high waste equipment, they discovered 98 mix syphons they didn't think they actually ensured over three locations. They saved $147,000 by using this single setting focused evaluation at a cost of $1,500 per person.

12. Conclusion

The reason for this study was to analyze different region of the writing connected with the utilization of blockchain innovation in the medical services field. Other associated region of the request were investigated also, like medical services data protection and security. The utilization of blockchain innovation in the field of medical services is a generally new thought, with the most outstanding papers in this space showing up somewhere in the range of 2016 and 2019. Thus, early examination in the field principally determined wide wording. Following that, underlying endeavors have been made to adjust this innovation to medical services information trade, drug production network the executives, and patient checking frameworks.

We can see from the improvement of this study that blockchain innovation can be utilized in different routes in the medical services region. The IoT-empowered medical care hardware checking is one of them worth exploring further. Because of the entire assortment of shrewd wellbeing gadgets that are at present arising, it is interesting to talk about the capability of conveying more security to the wellness and mental medical care checking conditions with these advancements together.

A few issues influencing the trustworthiness and security of patient information could be relieved with the assistance of blockchain innovation. At long last, we summed up a few blockchain for medical care approaches and applications that are pertinent to every information region referenced in this review. The advantages and downsides of these systems to lay out a mark of correlation between them. For future examinations, a standard can be utilized.

References

[1] Melanie Swan. 2015. Blockchain: Blueprint for a New Economy (1st ed.). O'Reilly Media, Inc., Sebastopol, CA.

[2] Matthew B. Hoy. 2017. An introduction to the blockchain and its implications for libraries and medicine. Medical Reference Services Quarterly, 36(3): 273–279. DOI:https://doi.org/10.1080/02763869.2017.1332261.

[3] Thomas McGhin, Kim-Kwang Raymond Choo, Charles Zhechao Liu and Debiao He. 2019. Blockchain in healthcare applications: Research challenges and opportunities. Journal of Network and Computer Applications, 135(2019): 62– 75. DOI:https://doi.org/10.1016/j.jnca.2019.02.027.

[4] Onyejekwe, E.R. 2014. Big data in health informatics architecture. In 2014 IEEE/ACM International Conference on Advances in Social Networks Analysis and Mining (ASONAM'14). IEEE, 728–736. DOI:https://doi.org/10.1109/ASONAM. 2014.6921667.

[5] Jesse Yli-Huumo, Deokyoon Ko, Sujin Choi, Sooyong Park and Kari Smolander. 2016. Where is current research on blockchain technology?—A systematic review. PLoS One, 11(2016): 1–27. DOI:https://doi.org/10.1371/journal.pone. 0163477.

[6] Joshua Lind, Ittay Eyal, Florian Kelbert, Oded Naor, Peter R. Pietzuch and Emin Gün Sirer. 2017. Teechain: Scalable blockchain payments using trusted execution environments. CoRR abs/1707.05454 (2017). arxiv:1707.05454 http: //arxiv.org/abs/1707.05454.

[7] Cynthia Dwork and Aaron Roth. 2014. The algorithmic foundations of differential privacy. Foundations and Trends in Theoretical Computer Science, 9, 3–4 (Aug. 2014), 211–407. DOI:https://doi.org/10.1561/0400000042.

[8] Joshua Lind, Ittay Eyal, Florian Kelbert, Oded Naor, Peter R. Pietzuch and Emin Gün Sirer. 2017. Teechain: Scalable blockchain payments using trusted execution environments. CoRR abs/1707.05454 (2017). arxiv:1707.05454 http: //arxiv.org/abs/1707.05454.

[9] Sabt, M., Achemlal, M. and Bouabdallah, A. 2015. Trusted execution environment: What it is, and what it is not. In 2015 IEEE Trustcom/BigDataSE/ISPA, Vol. 1. IEEE, 57–64. DOI:https://doi.org/10.1109/Trustcom.2015.357.

[10] U.S. Department of Health & Human Services (HHS). 2013. Summary of the HIPAA Privacy Rule. Retrieved February 4, 2019 from https://www.hhs.gov/hipaa/for-professionals/privacy/laws-regulations/index.html.

[11] Abbas Acar, Hidayet Aksu, A. Selcuk Uluagac and Mauro Conti. 2018. A survey on homomorphic encryption schemes: Theory and implementation. ACM Computing Surveys 51, 4 (July 2018), Article 79, 35 pages. DOI:https: //doi.org/10.1145/3214303.

[12] Eli Ben-Sasson, Alessandro Chiesa, Christina Garman, Matthew Green, Ian Miers, Eran Tromer and Virza Madars. 2014. Zerocash: Decentralized Anonymous Payments from Bitcoin (extended version). Retrieved February 5, 2018 from http://zerocash-project.org/media/pdf/zerocash-extended-20140518.pdf.

[13] Bethencourt, J., Sahai, A. and Waters, B. 2007. Ciphertext-policy attribute-based encryption. In IEEE Symposium on Security and Privacy (SP'07). IEEE, 321–334. DOI:https://doi.org/10.1109/SP.2007.11.

[14] Karafiloski, E. and Mishev, A. 2017. Blockchain solutions for big data challenges: A literature review. In 17th International Conference on Smart Technologies (IEEE EUROCON'17). IEEE, 763–768. DOI:https://doi.org/10.1109/ EUROCON.2017.8011213.

[15] Douglas M. Lambert, Martha C. Coope, and Janus D. Pagh. 1998. Supply chain management: Implementation issues and research opportunities. The International Journal of Logistics Management 9(2): 1–20. DOI:https://doi.org/ 10.1108/09574099810805807 arXiv:https://doi.org/10.1108/0957409981080580.

[16] Khezr, S., Moniruzzaman, M., Yassine, A. and Benlamri, R. 2019. Blockchain technology in healthcare: A comprehensive review and directions for future research. Appl. Sci., 9(9): 1736.

[17] T. Le Nguyen, T. 2018. Blockchain in healthcare: A new technology benefit for both patients and doctors. In 2018 Portland International Conference on Management of Engineering and Technology (PICMET), IEEE (2018 Aug 19), pp. 1–6.

[18] Khatoon, A. 2020. A blockchain-based innovative contract system for healthcare management Electronics, 9(1): 94.

[19] Jiang, S., Cao, J., Wu, H., Yang, Y. and Ma, M. 2018. He Blochie: A blockchain-based platform for healthcare information exchange. 2018 IEEE International Conference on Smart Computing (Smart Comp), IEEE (2018 Jun 18), pp. 49–56.

[20] Javaid, M. and Halee, A. 2019. Industry 4.0 applications in medical field: A brief review. Current Medicine Research and Practice, 9(3): 102–109.

[21] Dudovskiy, A. et al. 2017. How blockchain could empower eHealth: An application for radiation oncology VLDB Workshop on Data Management and Analytics for Medicine and Healthcare, Springer (2017).

[22] Blockchain based distributed control system for edge computing. Control Systems and Computer Science (CSCS), 2017 21st International Conference on (2017) IEEE.

[23] Deloitte. Blockchain in commercial real estate. The future is here! 2017 McFarlane C, Beer M, Brown J, Prendergast N. Patentor: a healthcare peer-to-peer emir storage network v1.0. 2017.

[24] Roma, P., Quarrel, F., Israel, A. et al. 2016. Blockchain: An enabler for life sciences and healthcare blockchain: An enabler for life sciences healthcare.Chronicled, Chronicled - Supply Chain Compliance. 2017. IBM Institute for Business Value. Healthcare Rallies for Blockchains, 2016.

[25] Mettler, Matthias. 2016. Blockchain technology in healthcare: The revolution starts here. 1–3. 10.1109/Healthcom.2016.7749510.

Chapter 5

Blockchain and IoT for Health-Care Applications

*Jasmin Annie Genefer M** and *Janeela Theresa MM*

Blockchain is a decentralized and public ledger that keeps track of transactions across a network of computers. It's a system that permits a group of untrustworthy individuals to verify a transaction. IoT devices in healthcare can give real-time sensory data from patients, which can be processed and evaluated.

The combination of IoT and Blockchain could be a viable option for designing a decentralized IoT-based healthcare system. The perfect combination of blockchain and IoT appears to be flourishing together, as there is a huge requirement for data security for the massive amounts of data created by IoT sensors.

Blockchain is a relatively advanced technology that stores data in a distributed manner and enhances security. This chapter proposes an integrated distributed application, a data protection system focused on an IoT framework and blockchain technology. The uses of Blockchain technology will play a vital role in keeping medical data preserved and transparent.

It is suggested to provide a path for the direction of future research, and to discuss new challenges associated with block chain-based healthcare data security implementation.

St. Xavier's Catholic College of Engineering, Nagercoil, Tamil Nadu, India.
Email: theresajaneela@yahoo.com
* Corresponding author: anniegenefer@sxcce.edu.in

1. Introduction

Blockchain is a decentralized and public digital ledger that records transactions on numerous computers in such a way that no record can be changed retrospectively without affecting subsequent blocks. Because any transaction is registered and checked publicly, each 'block' in the blockchain is confirmed and linked to the one before it, producing a long chain. Blockchain ensures a high level of transparency.

Blockchain is a distributed ledger network that adds data without requiring a consensus and never deletes or edits them. The cryptographic hash that connects freshly added information block records with each data block determines the hash value of a blockchain.

The underlying information (IT) architecture of the blockchain and its unbreakable chain of data entries that enable for secure and open transactions is the blockchain's main promise. Sharing medical data is a critical step in making the medical system more intelligent and improving medical service quality.

2. Blockchain and IoT Technology in Healthcare

The primary method for addressing health-care applications is remote patient monitoring. Data is collected and transferred to hospitals via wearable devices. In remote patient monitoring, IoT devices play a critical role. The major goal of these devices is to offer vital information to health experts, such as breathing patterns, blood glucose levels, and blood pressure [1].

Health devices that are used for data collection can be categorized into four parts:

Stationary medical devices: Devices are used for specific physical locations.

Medical embedded devices: Devices are placed inside the human body.

Medical wearable devices: Devices are medical wearable devices prescribed by the Doctors.

Wearable health monitoring devices: Devices were worned on the body.

Figure 1 shows IoT based health care system in blockchain, which are discussed below.

For the security of the patient's information, healthcare is an IoT system application that demands interoperability and data transfer. The process of sharing data with different sources is referred to as interoperability. The limitation to accomplish interoperability is part of the centralized notion. The Internet of Things is predicated on centralization, with data

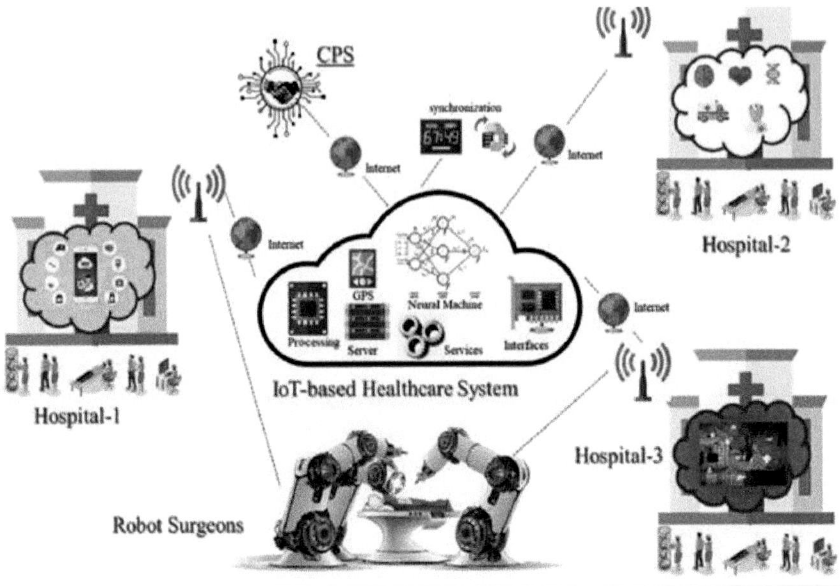

Fig. 1. IoT-based healthcare system.

being stored in the cloud, which is not safe. Blockchain combined with IoT can address the security challenges that healthcare applications meet [2].

2.1 Healthcare in IoT Applications

Some of the IoT applications related to healthcare using blockchain are listed below.

2.1.1 Master Patient Identifier (MPI)

Interoperability and autonomous automatic adjudication MPIs are single-person identities that can track a patient in a variety of settings, allowing for more seamless and scalable health care delivery across the continuum of care.

Autonomous automatic adjudication would streamline and improve the way clients or other healthcare transactions are handled between parties. The technique would involve a cross-party smart contract that would allow for automatic claim adjudication.

Interoperability between systems and organizations could also be improved with blockchain. Blockchain transactions would also benefit from being cryptographically and irreversibly secure, preserving anonymity for all parties involved.

2.1.2 Patient Monitoring/Electronic Health Record (EHR)

According to the International Organization for Standardization, an electronic health record is a digital record of a patient's data that is securely transferred and only accessible by authorized personnel. There are numerous EHR systems that use block chain technology.

Medrec: It is a block chain model used for authentication confidentiality and data sharing.

Gem Health Network: This network is used to replace the concept of data centralization with that of decentralization. Gem's health network framework is entirely built on Gem's fundamental feature of decentralisation. All of the records in the network are transparent, and any changes to the records will be reflected in all of the network's users.

Health Bank: It's a platform that keeps and secures health information. It's a new start-up that also offers some compensation to patients for their contributions.

Omni PHR: Public Health Record provides the facility for patients to assess their data. This model is developed to update the record and to differentiate between electronic health.

2.1.3 Healthcare Data Management in Blockchain

Health monitoring is another IoT device or networked device application, and data sharing and gathering should take place in a safe setting.

Information technology provided the opportunity to alleviate such effort by introducing EHRs. Furthermore, EHR allows for better disease control and more preventive care. In addition, the digital record improves decision-making capabilities and fosters greater collaboration among caregivers. As a result, the healthcare community is becoming more aware of its importance [3].

Figure 2 depicts the seven processes of a blockchain-based healthcare data management workflow, which are explained further below.

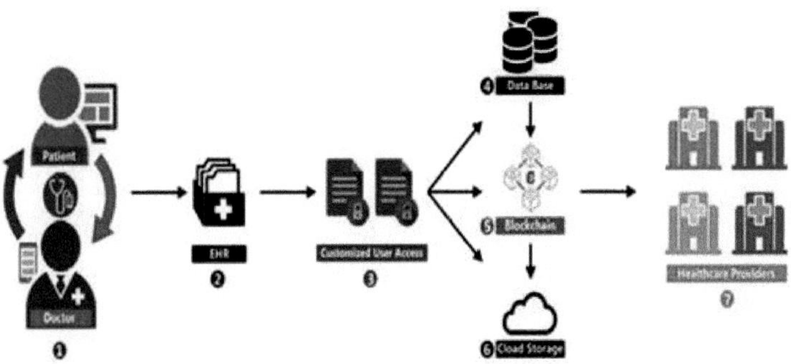

Fig. 2. Healthcare data management.

Step-1. The interaction between a patient and their doctors and specialists generates primary data. This information includes medical history, present symptoms, and other physiological data.

Step-2. Using the primary data acquired in the first stage, an EHR is constructed for each patient. Other medical data, such as that generated by nursing care, medical imaging, and prescription history, is also stored in the EHR.

Step-3. Individual patient who has the ownership of sensitive EHR, and customized is access controls given only to the owner of this property. Parties who want to access such valuable information must request permission which is forwarded to the EHR owner, and the owner will decide to whom access will be granted.

Step 4, 5, and 6 are part of the core of the whole process including database, the blockchain, and cloud storage. Database and cloud storage store the records in a distributed manner and a blockchain provides extreme privacy to ensure customized authentic user access.

Step-7. Healthcare providers such as ad hoc clinic, community care center, hospitals are the end user who wants to get access for a safe and sound care delivery which will be authorized by the owner. For example, no matter where you are treated in the globe, your health record will be available and accessible on your phone and validated through a distributed ledger such as blockchain, to which healthcare providers would continue to add to over time [3].

2.2 *Comparative Analysis of Blockchain and IoT in Healthcare*

Table 1 provides a comparison of blockchain with IoT smart contracts in healthcare system.

Table 1. Comparison between blockchain and IoT Smart Contracts.

Features	**Block chain**	**IoT Smart Contracts**
Network structure	Decentralization	Centralization
Transparency	It is high, although it can be limited to the application level	Low is determined by the contract's accessibility definitions for specific node classes
Cost	Low	High
Security	Very Secure	Less Secure
Scalability	Highly scalable	Low
Bandwidth	High Bandwidth	Nodes have limited
Latency	Needs time for information processing validating and mining	Demands lower latency

3. Platforms used for Blockchain

Blockchain has been highlighted as a disruptive technology that has the potential to have a significant impact on a variety of businesses, including the most popular and appropriate IoT areas [8].

1. **Ethereum:** Ethereum is a blockchain-based distributed computing platform and operating system with smart contract capability that is open-source and available to the public. It uses transaction-based state transitions to support a modified form of Nakamoto consensus.
2. **Bitcoin:** Bitcoin is the first cryptocurrency and an open-source international currency. It is the first digital currency to be distributed. We can send and receive Bitcoin from anywhere in the world, and we can use Bitcoin to buy things.
3. **Ripple:** Ripple is a remittance network, currency exchange, and real-time gross settlement system. It is based on a distributed open-source protocol and enables tokens that represent fiat cash, cryptocurrency, commodities, or other value units like frequent flier miles or cell minutes.
4. **Quorum:** Quorum is appropriate for any application that requires high-throughput processing of private transactions inside a permissioned group of known participants at a high pace. Quorum tackles specific issues with blockchain technology adoption in the banking sector and beyond.
5. **Hyperledger Sawtooth:** Sawtooth is a modular framework for developing, deploying, and running distributed ledgers. Distributed ledgers keep track of digital information (such as asset ownership) without the need for a central authority or implementation.
6. **Hyperledger Fabric:** The Linux Foundation hosts Hyperledger Fabric, a blockchain framework implementation. Hyperledger Fabric is a plug-and-play platform for constructing modular applications and solutions. Its components, such as consensus and membership services, are plug-and-play. Hyper ledger fabric makes use of container technology to host smart contracts known as chain code, which make up the system's application logic. As a result of the first hackathon, Digital Asset and IBM first contributed Hyperledger Fabric.
7. **HyperledgerIroha:** The Linux Foundation hosts Hyperledger Iroha, a blockchain platform implementation. Hyperledger Iroha is written in C++ and includes the BFT ordering service and a chain-based Byzantine fault-tolerant consensus method named yet another Consensus.

4. Integrating IoT with Blockchain for Healthcare

To safeguard medical data, a communication protocol between nodes based on the subscribe model, which is a model used exclusively by IoT devices, is being implemented. They create several smart contracts for IoT data producers and subscribers, as well as data owners' access permissions [10].

Figure 3 illustrate a blockchain based IoT system for health care applications.

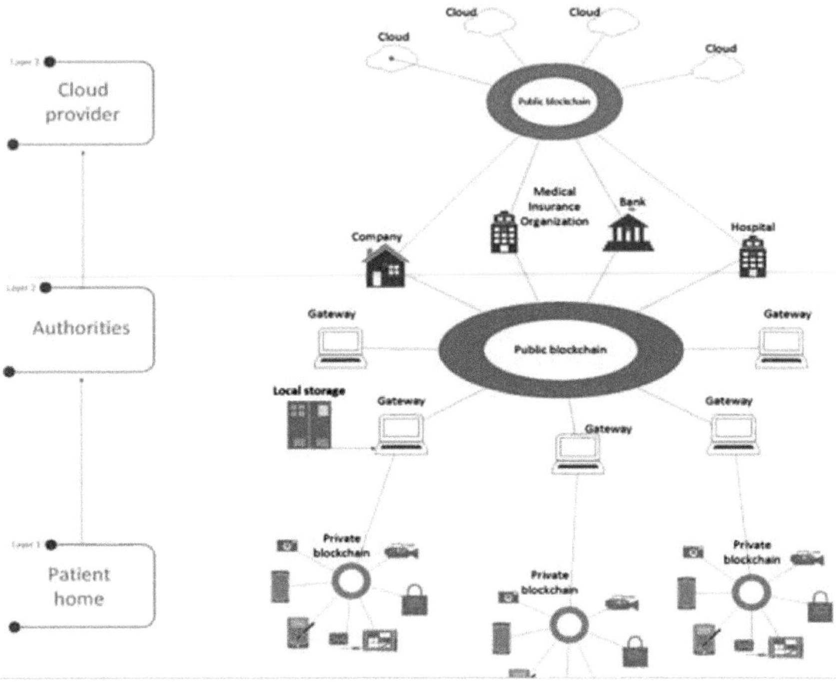

Fig. 3. Integrating of Blockchain with IoT [10].

Integrating IoT with Blockchain contains three layers

i) Patient Home
ii) Authorities
iii) Cloud Provider

i). Patient Home

It comprises all of the IoT nodes that are collecting data from a single patient. In this blockchain, a main node is a powerful computer that serves as a gateway to higher tier blockchains. Each patient will be assigned to their own blockchain.

ii) Authorities

It has representative nodes for all interacting medical parties with an interest in patient data, such as hospitals, medical centers, labs, and so on. This block chain also includes the gateways in the first layer. The authorities are the subscribers who are able to access the data of the publishers in the cloud.

iii) Cloud Provider

Processing and storage capacities in the cloud are typically used to complete IoT devices. A blockchain at the cloud level is required to govern the relationship between cloud providers and to provide access to patient data to each other, regardless of where it is stored.

Publishers who create all the data for a certain patient are known as gateways (medical data). Publishers define who has access to their data in the cloud and who has authorization to read, write, and alter it.

5. Scenario for Healthcare Applications

Scenarios comprised for healthcare application were discussed below:

Scenario 1 Authority wants to access patient's medical record

The records comprising the patient's data have already been stored in the patient's cloud provider in this case. The authorities with approval require access to a previously stored record. The use of a cloud contract is used to regulate access and permission. Using its account, the patient gets direct access to the data in the cloud. The cloud returns the ID of the record after it has been stored. The cloud then creates a transaction to indicate that the patient accessed that ID's record on this date. The authority transmits its ID to the correct cloud and awaits an acknowledgement. In this scenario, the ACK is the result of determining that the authority's ID is invalid [12].

Scenario 2 The gateway collects IoT data and generates a new record

The entire internet will be active in this circumstance. The sensors and their gateway are nodes in the private BC (BlockChain), therefore the miner is the gateway because it is the most powerful node in its private blockchain. Before sending data, every device must authenticate with the network using public and private keys. Each gadget has its own set of two keys. The gateway saves all of the keys in its local storage so that any device that authenticates with it can be quickly identified [12].

Scenario 3 Patient visits and Interact with an Authority

A new block is added to the layer 2 blockchain when a patient visits an authority. Consequently, the same block will be added to all authorities and a related block will be added to the cloud's provider blockchain. When the patient finishes its visit, the authority adds a new block to the public blockchain that includes the ID of the authority, the ID of the patient and some information about the data stored in the off-chain storage of this authority. The visited authority creates in the cloud's blockchain a block to note that the patient of this ID has visited the appropriate authority. It even notes the place where the data had been stored [12].

6. Challenges Addressed in Healthcare System using Blockchain

Many challenges are faced by the blockchain and it's due to the advancement of technology in health care.

Interoperability

Ensure adequate interoperability among them might be a challenge. Healthcare interoperability involves exchanging information with each other in a block chain network that can be diverse players like Hospital Insurance company physical private doctors Excel Crop [2].

Security

Due to security concerns, many patients may be hesitant to provide their personal medical information. One of the most significant issues about blockchain is security. Because blockchain applications are connected and accessible over the Internet, they are vulnerable to a variety of cyber-attacks, such as data theft, espionage efforts, and denial-of-service (DoS) attacks, which can render block chain services unusable [5].

Scalability

With the increasing expansion in the size and number of entities involved and transactions being completed, scalability has become a critical concern. Consider the fact that the rate of transactions that must be recorded in a blockchain system may be larger than the system's data synchronization capacity rate [4].

Lack of Standardization

Block chain is a popular technology that has been used in a number of countries. The Blockchain is employed in areas and networks where the concepts of security, trust, and traceability are important. Standardization of protocols, technology, and other elements is critical. What data, size, and format can be delivered to the blockchain, and what data can be stored on the blockchain are all factors to consider [5].

Data Ownership and Accountability

Other hurdles in integrating Block chain and IoT technology in the healthcare sector include data ownership and responsibility. The primary questions are who will hold the data, who will grant authorization to share people's private health-related data, and who will own it [6].

Table 2 shows the survey on blockchain solutions.

Table 2. Blockchain-based security solution in healthcare systems.

Reference	Issues	Blockchain solution	Advantage
[15]	Access Control	Ethereum based Smart contracts	Integrity, scalability and data ownership
[16]	Privacy issues	Private blockchain and consortium blockchain	Access control, auditing, secure search
[17]	Real-time patient monitoring security	Permissioned and consortium managed blockchain	Traceability, privacy and transparency
[18]	Data sharing, scalability and QoS	Block chain based smart and secure health-care system	Data sharing, QoS, efficient, remote monitoring
[19]	Denial of service, data sniffing attack	RHM solution based on a public blockchain	Privacy preserving, efficient message delivery times
[20]	Data leakage	Customized search index-based blockchain	Reliable search results, confidentiality
[21]	Privacy and anonymous information sharing issues	A new consensus algorithm.	Privacy protection and efficient transaction handling

7. Case Study

7.1 Patient-Centric System using Blockchain

This case study is to demonstrate how blockchain could be used in the e-healthcare industry to protect the security and privacy of EHR (Electronic Health Records). Figure 4 shows patient-centric system using blockchain. More information about this case study may be found at [14].

The case study employs a smart contract on the Ethereum consortium blockchain to move patient health records from the healthcare industry's management and control to a patient-centric application. Patients have control over their health data with this method.

In addition to patients and hospitals, the administrator is a user-end entity. The registration of hospital entities in the blockchain framework is the responsibility of this entity. Using the web application in the front end, hospitals enrolled with the blockchain framework could safely transfer patient data with the suggested patient-centric system. Patients have control of their data once it has been shared with the system, and they can share it with any other doctors or hospital entities who have registered with the Ethereum blockchain patient-centric system. This information is kept on a distributed ledger.

The system's performance and security are evaluated in [14]. It was demonstrated that by leveraging the Ethereum blockchain, security and performance could be guaranteed. The time it took to access such data was less than one second.

Figure 4 show the patient-centric system were discussed below.

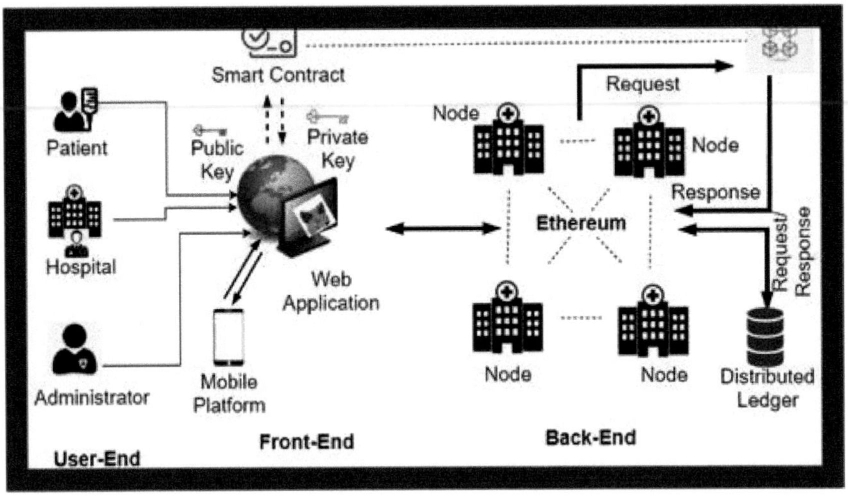

Fig. 4. Patient-centric system using blockchain [14].

8. Conclusion

This chapter describes the various applications of healthcare and how the block chain and IoT is integrated with healthcare. It also deals with Blockchain technologies, the database management in healthcare and comparison between blockchain and IoT. Blockchain platform for healthcare, integrating the IoT and Blockchain in healthcare and case study are also mentioned.

References

[1] Naresh, V.-S. 2020. Internet of things in healthcare: Architecture, applications, challenges, and solutions. Computer Systems Science and Engineering, 35(6): 411–421.

[2] McGhin, T., Choo, K.-K.R., Liu, C.Z. and He, D. 2019. Blockchain in healthcare applications: Research challenges and opportunities. Journal of Network and Computer Applications, 135: 62–75.

[3] Dimitrov, D.V. 2019. Blockchain Applications for Healthcare Data Management. Healthc. Inform. Res. 2019.

[4] Porru, S., Pinna, A., Marchesi, M. and Tonelli, R. 2017. Blockchain-oriented software engineering: Challenges and new directions. In Proc. 39th Int. Conf. Softw. Eng. Companion, 2017.

[5] 51% Attack, Web Article. Accessed: Jan. 2019. [Online]. Available: https://learncryptography.com/cryptocurrency/51-attack.

[6] Kumar, N.M. and Mallick, P.K. 2018. Blockchain technology for security issues and challenges in IoT. Procedia Computer Science, 132: 1815–1823.

[7] Cheng, J., Li, J., Xiong, N., Chen, M., Guo, H. and Yao, X. 2020. Lightweight mobile clients privacy protection using trusted execution environments for blockchain. Computers, Materials & Continua, 65(3): 2247–2262.

[8] Blockchain for IoT-Based Healthcare: Background, Consensus, Platforms, and Use Case. Partha Pratim Ray, SMIEEE, Dinesh Dash, Debashis De, SMIEEE, Khaled Salah, January 2020 IEEE Systems Journal, pp. 1–10.

[9] Blockchain Technology in Healthcare: A Comprehensive Review and Directions for Future Research. Seyednima Khezr, Md Moniruzzaman.

[10] Nabil Rifi, NazimAgoulmine, Nada ChendebTaher and ElieRachkidi. 2017. Towards using blockchain technology for IoT data access protection. IEEE 17th International Conference on Ubiquitous Wireless Broadband (ICUWB), 2017.

[11] Nabil Rifi, NazimAgoulmine, Nada Chendeb Taher and ElieRachkidi. 2018. Blockchain technology: Is it a good candidate for securing IoT sensitive medical data? Wireless Communications and Mobile Computing Journal, 2018.

[12] Integrating Blockchain with IoT for a Secure Healthcare Digital System Nada Chendeb, Nour Khaled, Nazim Agoulmine.

[13] McGhin, T., Choo, K.-K.R., Liu, C.Z. and He, D. 2019. Blockchain in Healthcare Applications: Research Challenges and Opportunities.

[14] Towards a Blockchain Assisted Patient Owned System for Electronic Health Records Tomilayo Fatoku , Avishek Nag and Sachin Sharma.

[15] Dagher, G.G., Mohler, J., Milojkovic, M. and Marella, P.B. 2018. Ancile: Privacy-preserving framework for access control and interoperability of electronic health records using blockchain technology. Sustainable Cities and Society, 39: 283–297.

[16] Zhang, A. and Lin, X. 2018. Towards secure and privacy-preserving data sharing in e-health systems via consortium blockchain. Journal of Medical Systems, 42(8).

[17] Griggs, K.N., Ossipova, O., Kohlios, C.P., Baccarini, A.N., Howson, E.A. and Hayajneh, T. 2018. Healthcare blockchain system using smart contracts for secure automated remote patient monitoring. Journal of Medical Systems, 42(7).

[18] Abdellatif, A.A., Al-Marridi, A.Z., Mohamed, A., Erbad, A., Chiasserini, C.F. and Refaey, A. 2020. sshealth: Toward secure, blockchain-enabled health-care systems. IEEE Network.

[19] Ali, M.S., Vecchio, M., Putra, G.D., Kanhere, S.S. and Antonelli, F. 2020. A decentralized peer-to-peer remote health monitoring system. Sensors, 20(6).
[20] Chen, L., Lee, W.-K., Chang, C.-C., Choo, K.-K.R. and Zhang, N. 2019. Blockchain based searchable encryption for electronic health record sharing. Future Generation Computer Systems, 95.
[21] Du, M., Chen, Q., Chen, J. and Ma, X. 2020. An optimized consortium blockchain for medical information sharing. IEEE Transactions on Engineering Management.

Chapter 6

Secure E-Health System using Blockchain Technology in IoT Environment

Anitha Amaithi Rajan[1,*] and *Yamuna Bee J*[2]

E-Health is a sort of contemporary healthcare that benefits from the application of numerous communications technology and communicative capabilities. A large variety of patient information was managed in one location so that it may be transmitted elsewhere as needed. Such information must therefore be held in the tightest of confidentiality. Blockchain makes it possible for people who don't trust one other to converse with each other in a scientifically verifiable manner without the necessity of outside help. Moving on to the Internet of Things, we discuss how a blockchain-IoT aggregate that permits the sharing of offerings and sources creates a market of services among devices and enables us to automate many now, lengthy processes in a cryptographically provable way. This gadget described the relationship between healthcare data and the blockchain IoT network. The improved blockchain network will no longer simply satisfy healthcare needs, but will also enable more secure client engagement.

[1] Research Scholar, Department of Computer Science Engineering, Francis Xavier Engineering College, Tirunelveli, Tamil Nadu.
[2] Assistant Professor, PSN College of Engineering and Technology, Department of Computer Science Engineering, Tirunelveli, Tamil Nadu.
Email: yamikhan@gmail.com
* Corresponding author: anitharajan1804@gmail.com

1. Introduction

The improvement of health and fitness care services via the use of virtual technology and telecommunications, such as computers, the Internet, and mobile devices, is known as e-health care. E-health is commonly combined with conventional "off-line" (non-virtual) approaches to provide patients and consumers of fitness care with information. Demand for e-fitness is strong. E-fitness was created in response to the demand for more sophisticated patient tracking and documentation, especially for financial reasons like reimbursement from insurance companies. In the past, medical professionals appeared in front of respective patients and kept paper notes. On the other hand, rising fitness-related expenditures and technological improvements pushed for the development of digital monitoring systems. As e-fitness technology advanced, the area of telemedicine, which use communications technology to deliver healthcare—emerged.

1.1 *The Need for E-Health*

E-health emerged from the necessity for more sophisticated tracking and documentation of a patient's health and behavior, particularly for financial considerations, such as those made by insurance companies. In the past, medical professionals stood in front of their patients and kept paper notes. On the other hand, rising health care expenses and technological improvements pushed for the development of digital monitoring systems. As e-fitness technology advanced, the area of telemedicine—which use communications technology to deliver healthcare—emerged remotely.

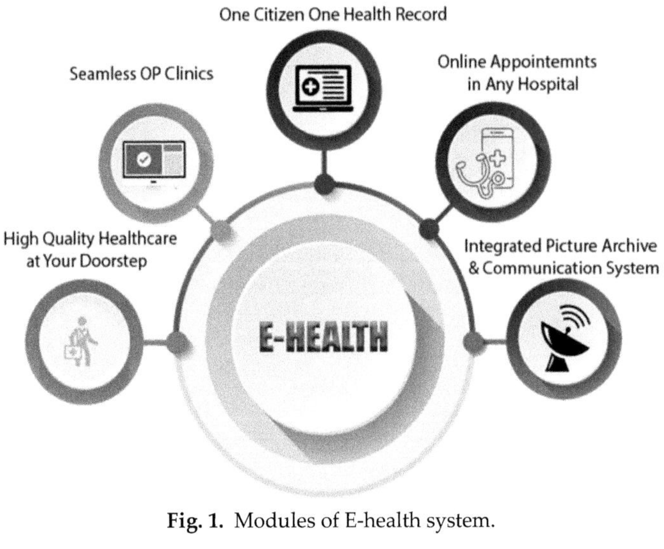

Fig. 1. Modules of E-health system.

1.2 E-Health Technology

Virtual technology is widely used in e-health. For instance, the Internet enables users of health services to contact medical professionals, obtain clinical data, assess fitness statistics, and have man-to-man or woman exchanges of text, audio, video, and other records. Interactive TV, also known as Polycom, enables two or more people in two or more places to communicate in real-time via voice and video. In e-health, freestanding devices called kiosks—typically PCs—are used to provide interactive information to users. A series of interactive activities on a suggestion screen serves as the main source of knowledge. Additionally, clients' data and information may be gathered through kiosks. On DVDs, USB flash drives, and other storage devices, digital recordings are preserved. Many contemporary mobile devices come with personal computing, Internet, and downloadable applications (or apps) that enable users to easily access medical data. All customers, even those with impairments like blindness or hearing, can use many of the technologies employed in e-fitness.

This e-Health device's main objective is to improve the next-generation healthcare system with virtual service solutions to decrease costs, improve patient outcomes, and secure, reliable, flexible, and efficient solutions to complicated e-Health challenges [1]. The security approach was used in the past with the use of password authentication techniques, wireless sensor systems that could communicate over small distances or a few other types that were simple targets for expert hackers. As a result, it is imperative to modify e-Health technical goods to fulfill people's demands, both in terms of sheer population and the development of social interaction norms.

This e-Health device's main objective is to improve the next-generation healthcare system with virtual service solutions to decrease costs, improve patient outcomes, and secure, reliable, flexible, and efficient solutions to complicated e-Health challenges [1]. The security approach was used in the past with the use of password authentication techniques, wireless sensor systems that could communicate over small distances or a few other types that were simple targets for expert hackers. As a result, it is imperative to modify e-Health technical goods to fulfill people's demands, both in terms of sheer population and the development of social interaction norms.

1.3 The Future of Blockchain in Healthcare

One of the most used keywords in the scientific community right now is "blockchain." This is sole because of this. Simply said, blockchain technology has the power to revolutionize healthcare. With its full implementation, patients may be precisely targeted in the middle of all procedures, which may then be fully rebuilt with increased security, privacy, and accessibility. However, how can blockchain enable all of this? How is the fitness sector making use of this extremely prosperous modern age?

1.4 What is Blockchain?

We should investigate what blockchain is. A disseminated machine that creates and stores records is known as a blockchain. On its distributed organization, it keeps a virtual record of connected "blocks" of insights that address how records are shared, altered, or got to.

At the point when a connected instrument is engaged with an exchange, all gadgets on the equivalent blockchain machine will create equivalent blocks. On the off chance that the records on one PC are gotten to, altered, shared, or generally changed in any capacity, a block is made to report the data on each device locally. Along these lines, changes to records can be made without causing challenges. A decentralized method permits records equality to be accomplished by evaluating the blocks of each connected device. The utilization of "hashing" is at the pinnacle of plainly recording and examining records. Hashing relegates each block to a totally special identifier that changes relying upon its items. If the records in a block substitute, the hash should.

This is significant since blocks are saved in ordered succession and reference the earlier block's hash. Thus, endeavoring to trade the records of one block may now be the impetus for a resulting block to distinguish a hash trade. It is because of this that connected blocks structure a "chain" that is unchangeable, reliable, and decentralized. Blockchain makes it almost difficult to copy, control, or in any case create records since it is precise, predictable, and an impetus for the greatest obligation. This opens up a few prospects, the most significant of which is the trade, stockpiling, and admittance to records among connected parties.

1.5 How Blockchain is Affecting the Medical Industry

Blockchain is an open and incredibly secure technology that may be used in a variety of ways in the medical sector, most notably to significantly reduce costs and create new methods for patients to get care. Future-proof technology can create a generation of growth and innovation when paired with the compounding nature of records and invention. A blockchain revolution is already being laid out by top organizations. The medical industry is secured by blockchain, as seen in the following figure.

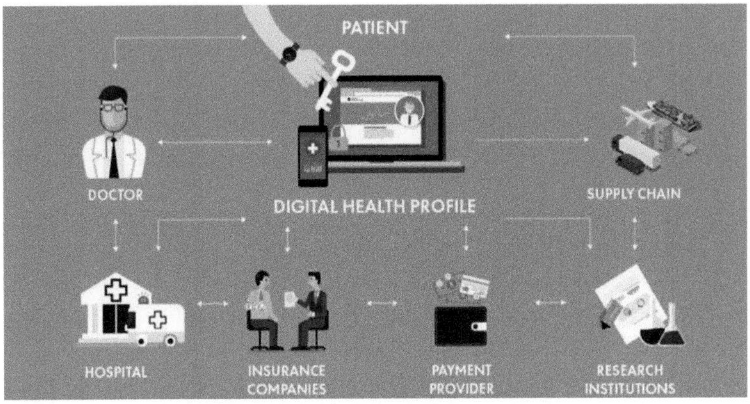

Fig. 2. Blockchain in medical industry.

2. Blockchain Network Technology

2.1 Working of Blockchain

- The most generally utilized sort of blockchain is to copy and share a portion of the local area's individuals. To take care of the twofold spending issue, this arrangement [2] has been utilized rather than the Bitcoin procedure. Subsequently, the Bitcoin blockchain fills in as an authoritative data set of exchanges, figuring out who claims what. The blockchain, then again, is a status no matter what anyone else might think, and there is no lack of bitcoin. In the blockchain, the measurements are assembled into timestamped blocks. Each block alludes to the hash of the block that preceded it, and each block is recognized by its cryptographic hash. This makes a connection between the blocks, which is alluded to as a chain of blocks or blockchain (Fig. 3). Any hub with admittance to this arranged, returned-related posting of blocks can look at it and make sense of the realm of measurements being changed over locally. We can acquire a superior comprehension of blockchain by concentrating on how blockchain networks work. A blockchain local area is a gathering of fixed hubs or clients that all work on the equivalent blockchain, with a duplicate of insights, hung on every hub. These individuals structure a shared organization in which:
  - Clients cooperate with a blockchain using private and public keys, which are utilized for confirmation. A client's hub communicates the marked exchange to its one-jump mates through the private/public keys.

- Before proliferating this approaching exchange any farther, the neighbors check that it is genuine; invalid exchanges are dismissed. This exchange in the end spreads all through the whole town.
- The exchanges that have been totaled and laid out involving the procedure above at a specific time in the C programming language are coordinated and bundled into a timestamped up-and-comer block. Whether the hash is made over the items in the block or its header, as in Bitcoin, is a format decision. The arrangement can be the overall population key itself or (regularly) a hash of it known as mining, contingent upon the execution. The mining hub declares that the block has been gotten back to the local area.
- The hubs guarantee that the proposed block (a) contains substantial exchanges and (b) references the ideal going before the block on their chain utilizing hash. If so, they add the block to their chain and track the exchanges it contains to keep their worldwide view up to current. The proposed block is wiped out on the off chance that this isn't generally the situation. This closes the series of adjustments.

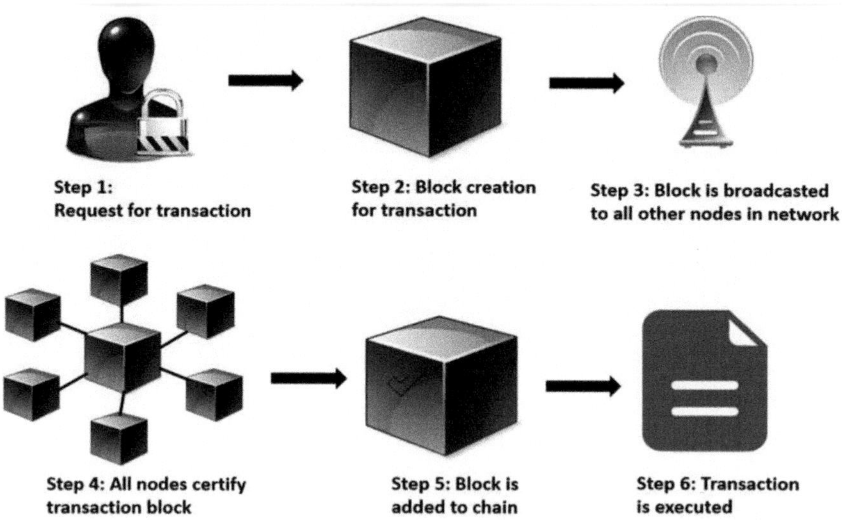

Fig. 3. Working of blockchain.

2.2 *Transferring Digital Asset Son a Blockchain*

With each mined block on account of Bitcoin, a new Bitcoin is given to the local area: Within the block of exchanges it broadcasts to the local area, and the mining hub contains a purported coin base exchange. This coin base exchange has no sources of info and prizes the mining hub with a predefined measure of bitcoins (chosen by the local area). The significant thing to recollect is that assuming that you have a gathering of clients that (a) need to trade virtual tokens and (b) settle on how those tokens are produced, then a blockchain local area is an ideal instrument to utilize for both supplanting tokens and monitoring who has what. There is no requirement for a go-between to empower the exchange rationale; every hub locally plays out the important tests and settles on the standard result. Resource observing is implicit since every hub has a passage in the blockchain settled upon the set of cryptographically unquestionable exchanges.

2.3 *IoT in Healthcare*

Before the Internet of Things, only in-person visits, teleconferences, and text messaging could patients communicate with their physicians. Doctors and hospitals were unable to regularly monitor their patient's health and provide guidance.

2.3.1 *Internet of Things (IoT)*

Associated gadgets have empowered a sweeping following inside the medical services industry, opening the capacity to keep patients protected and sound while additionally enabling doctors to give unrivaled therapy. It has likewise expanded patient commitment and fulfillment as collaborations with specialists have become simpler and more productive. Besides, remote observing of a patient's wellbeing supports diminishing the length of stay in the medical clinic and forestalling re-confirmations. IoT likewise altogether affects essentially reducing medical services expenses and further developing therapy results. IoT is changing the medical care industry by reexamining the distance among gadgets and how they associate to create medical care arrangements. In medical services, IoT has applications that benefit patients, families, doctors, emergency clinics, and insurance agencies.

IoT for Patients - fitness bands and other wirelessly connected wearables like glucometers, blood pressure, heart rate monitors, etc., give each patient specialized treatment. These gadgets may be set up to remind you of a variety of activities, including blood pressure checks, exercises, appointments, and much more. IoT has revolutionized people's life by enabling ongoing health monitoring, especially for older people. The individuals who live with me and their families are significantly impacted by this. The warning system warns a person's circle of contributors and fitness providers when their regular sports are interfered with or altered.

IoT for Physicians - Doctors can keep a superior check of their patients' wellness by utilizing wearables and other IoT-empowered family GPS beacons. They can screen patients' adherence to therapy suggestions and any requirement for sure-fire clinical help. IoT empowers medical services experts to be more cautious and proactive in their collaborations with patients. Information gathered from IoT gadgets can help doctors in deciding the best treatment technique for their patients and accomplishing the ideal results.

IoT for Hospitals - IoT devices are quite beneficial in hospitals for a variety of additional purposes in addition to monitoring patient health. Sensor-enabled Clinical equipment including wheelchairs, defibrillators, nebulizers, oxygen pumps, and other tracking devices are all tracked via IoT devices. Additionally, real-time monitoring of the placement of medical workers in out-of-the-ordinary areas is possible. Patients in hospitals are often concerned about the spread of infections. IoT-enabled hygiene monitoring tools aid in preventing patient infection. Additionally, IoT devices assist with asset management, such as pharmacy stock monitoring, and environmental monitoring, such as checking fridge temperatures and regulating temperature and humidity.

IoT for Health Insurance Companies – With IoT-associated brilliant gadgets, wellness guarantors have various choices. Insurance agencies can utilize the information accumulated from wellness trackers to work on their endorsing and asserts processes. This data will permit them to detect extortion guarantees and distinguish planned endorsing potential open doors. Inside the guaranteeing, estimating, claims the executives, and hazard assessment methodology, IoT gadgets give straightforwardness among safety net providers and clients. Clients could have adequate permeability into the hidden ideas driving every choice made and the way implications in the light of IoT-caught data-driven determinations in each activity methodologies.

Guarantors may likewise offer motivations to their clients for utilizing and sharing wellness information got by IoT gadgets. They can compliment clients on their utilization of IoT gadgets to keep music in their normal games, as well as their adherence to treatment regimens and safeguard wellness measures. This will significantly help back up plans in lessening claims. IoT gadgets can likewise help insurance agencies approve claims by utilizing information gathered by the gadgets.

2.3.2 Redefining Healthcare

The development of IoT products with precise healthcare functionality opens up enormous opportunities. Furthermore, the enormous amount of data generated by those linked devices continues to have the potential to revolutionize healthcare.

IoT has a 4-step structure that resembles degrees in certain ways (see Fig. 1). The four degrees are linked together such that information is acquired or processed at one degree and produces the price at the following degree. Integrating values into a process generates insights and offers exciting commercial opportunities.

Step 1: The initial step involves the arrangement of organized gadgets, like computerized cameras, sensors, actuators, screens, and indicators. These gadgets gather information.

Step 2: Typically, information gathered from sensors and different gadgets is in a simple structure, which should be consolidated and changed into a virtual structure for equivalent information handling.

Step 3: After the information has been digitized and gathered, it is pre-handled, normalized, and moved to the server farm or Cloud.

Step 4: The ideal measure of control and examination is applied to the last data. When exceptional examinations are applied to this information, it produces valuable business bits of knowledge for compelling independent direction.

IoT is reforming medical services by guaranteeing better consideration, high-level therapy results, and lower costs for patients, as well as better techniques and work processes, high-level execution, and a superior patient encounter for medical services experts.

2.3.3 The Fundamental Blessings of Iot in Healthcare Include

Cost Reduction: IoT makes it possible to follow patients in real-time, thereby reducing the number of unnecessary doctor visits, hospital stays, and re-admissions.

Improved Treatment: It gives complete openness and enables doctors to make decisions that are informed by evidence.

Faster Disease Diagnosis: Continuous patient monitoring and real-time data help in making early diagnoses of diseases, maybe even before they manifest themselves based just on symptoms.

Proactive Treatment: Continuous fitness monitoring makes it possible to provide proactive treatment.

Drugs and Equipment Management: A top-notch endeavor in the healthcare industry is the management of medications and medical equipment. These are managed and applied effectively at lower costs thanks to connected devices.

Error Reduction: Data created by IoT devices now not only aid in effective decision making but also ensure clean healthcare operations with reduced mistakes, waste, and device costs.

IoT in healthcare isn't always without issues. Concerns concerning information security are growing as a result of the IoT's connected devices capturing large amounts of sensitive data. It is essential to include adequate security features. Through real-time fitness monitoring and access to patients' medical records, IoT explores new facets of patient care. For healthcare workers looking to increase revenue, enhance patient care, and enhance patient tales, this information is a gold mine. In a growing number of connected worlds, organizing to harness this virtual power may set you apart from the competition.

3. Cloud Computing in Health Care

The medical care area is convoluted since there are countless various cycles and delicate and confidential information that should be dealt with. The intricacy of the area habitually brings about two critical difficulties: rising functional expenses (counting information stockpiling costs) and the issue of establishing an independent well-being climate. Significant impediments in the medical care industry have forever been overwhelmed with the assistance of innovation. Distributed computing is one such advancement. It has been being used in the medical services area for some time and is continually changing to reflect improvements in the area. With abilities like coordinated effort, versatility, reachability, effectiveness, and security, distributed computing is altering the medical services area on different levels.

At the point when medical services organizations and care suppliers need to convey, recover, and oversee network data right away, the cloud's on-request processing capacity offers esteem. Medical services information age, use, further developed stockpiling, cooperation, and sharing practices need to change considering the expansion deprived of information-based security. It is the region where distributed computing plays it safe!

One industry that has been at the cutting edge of embracing cloud innovation is medical care. Around the world, medical care suppliers are starting to figure out the genuine capability of cloud arrangements. By 2022, industry members are anticipated to spend roughly 35 billion bucks overall on distributed computing, as indicated by a BBC concentrate on examination.

By 2025, the market for cloud-fueled medical care is supposed to be valued at around 55 billion bucks, with the CAGR of cloud administrations and arrangements proceeding with its vertical pattern of 15%.

This article explains the key ways distributed computing will influence the medical services area. Here is a choice of requests that might provoke your interest in becoming familiar with distributed computing answers for the medical care industry.

- ✓ The Function of Cloud Computing
- ✓ What Is a Cloud-Based Healthcare System?
- ✓ Why do healthcare organizations use the cloud?
- ✓ Problems with the Healthcare Sector Healthcare Systems based on the Cloud vs. the Past How do cloud-based solutions address the challenges facing the healthcare sector?
- ✓ What role does cloud computing play in the medical industry?
- ✓ What applications does the healthcare industry have for cloud computing?

3.1 What is Cloud Computing?

Expanding interest in innovation has constrained dependable foundation choices in IT and information stockpiling. The idea of cloud innovation has been standing out as of late. In its easiest structure, "distributed computing" alludes to the on-request accessibility of PC framework assets like information stockpiling and handling power. The expression "cloud" is utilized to portray a server farm that is accessible to numerous clients over the Internet and is conveyed across various areas from a focal server. Distributed computing makes it simple to store archives in a single spot that is open from any gadget whenever.

At first, all applications and programming must be on a PC or server that must be gotten to from one specific area. With the presentation of the cloud, individuals can get to their projects and data utilizing the Internet. This guideline additionally applies to information capacity. Organizers brimming with significant work are kept on your PC and server, however, you can store your information from a distance and back it up to the cloud.

For instance, Netflix is dealing with distributed computing to upgrade video web-based features and other business frameworks.

3.2 What is a Cloud-Based Healthcare System?

Cloud-based medical care alludes to the coordination of distributed computing advances for the creation and the board of cloud-based medical services administrations. Increasingly more medical services suppliers need to work with sellers that give distributed computing answers for putting away and recovering computerized records. This is viewed as a significant advantage for huge and little supplier associations, as data can be securely put away offsite. Cloud-based medical care frameworks meet the accompanying key necessities in the medical care industry:

On-request admittance to PCs with enormous capacity assets not accomplished with conventional clinical frameworks. Upholds huge datasets for offloading EHR, radiographic pictures, and genomic information. The capacity to impart EHRs to guaranteed doctors, emergency clinics, and care offices across geographic areas. This gives opportune admittance to lifesaving data and limits the requirement for copy testing. Further, develop examination and observation of demonstrative, remedial, cost, and execution information. Cloud-based medical services frameworks lessen functional expenses and convey better medical care administrations while giving more customized care and productive work processes. Simultaneously, patients can get quick reactions from their medical services suppliers, further develop a following and access their wellbeing records. On the off chance that the medical services environment is overseen on a neighborhood server, it will incorporate EMR and clinical charging framework. Accordingly, the costs will be high because of the above variables, for example,

- Support overhauling records
- Reinforcement conveniences
- Load adjusting issues
- Space Utilization

With these deficiencies, picking a cloud well-being technique is a need of time. Embracing a cloud-based wellbeing framework would deal with all overheads and foundations. The cloud-based medical services framework fosters a restricted arrangement in distant patient checking with telehealth and telemedicine arrangements. Controlling simple interoperability with a coordinated pecking order is the target of any cloud-based medical care framework. An efficient cloud-based medical services framework grows new bits of knowledge for medical care the board arrangements. As distributed computing is a goliath shared pool in the medical services industry, cloud arrangements can increase or psychologist all the stockpiling assets and adjust to consistently evolving needs.

3.3 *Why are Health Organizations moving to the Cloud?*

To work on persistent consideration, present creative patient consideration applications, and advance tasks, medical services organizations have been executing innovations. Disregarding these IT arrangements, they should battle with issues including high framework costs, the interest for processing assets, adaptability, general access, multi-tenure, and an expansion in coordinated effort needs. These challenges are tended to by the cloud's highlights, which include:

- On-request administration: Without any human contribution, the assets are conveyed immediately.
- Asset pooling empowers different clients to use cloud benefits simultaneously.
- Flexibility: It is attainable to add, erase, or redesign contingent upon the necessities of the organization.
- Expansive Network Access: From each area whenever a wide assortment of organization openness is advertised.
- Estimated administrations expect clients to just compensation for the administrations they use.

3.4 Traditional vs. Cloud-Based Health Care Systems

i. Customization

Before electronic health records, which allowed for customization, the desired bespoke solution had to be developed by highly qualified programmers and IT specialists. Contrarily, cloud-based systems have built-in functionality and care plans and are completely customizable. As a result, there are many different specialized templates and simple user interfaces available for cloud-based bespoke solutions.

ii. Potential Uses

Customary clinical frameworks require servers, inward information stockpiling, programming and equipment to be introduced in the specialist's field. Today, cloud-based frameworks give the comfort and flexibility to sign in from any area on any gadget through a web server. Likewise, basic interoperability, joint effort, and information sharing through cloud administrations have become more available. Cloud-based administrations permit programmed updates, upkeep, and information admittance to guarantee that your data matches the most recent form. Refreshing such data in customary medical care frameworks can be exceptionally complicated, tedious, and exorbitant.

iii. Compliance and Responsibility

While picking a customary or cloud-based medical care administration, taking into account unofficial laws and HIPPA requirements is insightful. Rules and guidelines need to address the requirement for safeguarded wellbeing data connected with HIPPA consistency, access control, and security concerns. With cloud-based medical services frameworks, care suppliers don't need to stress over framework blackouts, catastrophic events, and weather conditions that will generally crash the framework. Reinforcement prerequisites, code set refreshes, security patches, and conventions are unsure and unthinkable in customary medical services frameworks. Safeguarded wellbeing data in the cloud can be gotten to whenever anyplace. It is vital to take note that responsibility issues should constantly be HIPPA consistent.

IV. Wellbeing

Customary electronic servers will quite often be more powerless against malware, infections, and hacking endeavors than cloud-based servers, however, both require security alarms. Safeguarded Health Information (PHI) security is the main issue whether or not you pick a cloud-based medical care framework or a customary medical care framework. The two frameworks require standard examination, reviewing, and capacity limit checking. Human association with the two frameworks is fundamental because the utilization of cell phones builds the gamble of cyberattacks against PHI. With a cloud-based EHR framework, the execution of constrained encryption is more endlessly secure than conventional paper recording and client/server frameworks. This approach relies upon where these frameworks are and how they are safeguarded. Here is a portion of the medical services provokes that lead to the requirement for distributed computing in the medical services industry:

3.5 *Pain Points of the Healthcare Industry*

Here are the problems that the healthcare sector was facing before cloud computing.

1. Data Transfer Errors

Hospitals and other chained healthcare organizations experience problems with data transfer. Any medical facility handling an emergency is at risk because there is so much data to process and it takes so long. Cloud computing makes it easy and centralized to store data. This data may be accessible from the central data center concurrently from several places by avoiding flawed data transmission processes.

2. Slow Motion

When utilizing outdated data management systems, healthcare professionals regularly suffer speed issues. As a result, the system's slowness is to blame for its inability to handle a huge amount of data in the permitted time. As the volume of data increases, the speed decreases. There might be data loss as a result. Cloud computing in the healthcare industry is well known for its data accuracy and speed, nevertheless. They swiftly and continually process the data without pausing.

3. Records' Uncertainty

For medical organizations, managing scattered information from different sources is a significant test. It should be isolated and sorted for simple access. Since most offices are deficient, medical care experts can't get the fundamental information sooner rather than later. To deal with such circumstances quickly and helpfully, the medical care cloud assists with information capacity and support.

4. Inventory Problems

The healthcare facilities must keep their inventories current. However, probably, you won't remember the product's inventory. In certain multi-specialty hospitals, this might result in a scarcity of essential medical equipment and pharmaceuticals during medical emergencies. Since cloud computing can manage such crucial data successfully, issues with its acceptance in healthcare may never arise.

3.6 *How Can Cloud-Based Solutions Overcome the Pain Points of the Healthcare Industry?*

Distributed computing offers various advantages by making clients open to the framework, stages, and programming given by cloud suppliers. There are two sorts of distributed computing in medical services. Deals models (equipment/programming) can be presented as Software as a Service, Infrastructure as a Service, and Platform as a Service. The conveyance model (proprietor) can be private, cooperative, public, and mixture. The following are the advantages of imaginative cloud-based answers for beating the difficulties of the medical care industry.

1. Concentrated admittance to computerized patient records Previously, all patients had separate documents or clinical records for each specialist's visit. Keeping up with and overseeing desk work is quite difficult for specialists and staff. This cycle is currently being supplanted and is simpler to make do with cloud movement. With cloud benefits, all clinical records are in a solitary focal area. These records will keep on being available from the wellbeing Center web-based interface and can be recovered depending on the situation. A solid cloud stage ensures information capacity choices by facilitating arrangements and virtual machines for fast admittance to clinical records and speedy determination of patients.

2. Further developed patient consideration guidelines Cloud-based medical services arrangements work on understanding consideration by carrying imaginative treatments to the work area. With only a couple of

snaps, patients can begin virtual meetings, plan meetings with their PCPs, run programmed updates for impending arrangements, and track them through cloud administrations. After medical procedures, the cloud stage gives normal connections to specialists, drug updates, and data about forthcoming offices. Specialists can likewise remotely screen the patient's important bodily functions with the assistance of a clinical imbuement siphon associated with the cloud. Clients can share, view, and store their clinical records in the cloud, however, specialists can likewise remotely file and access them. Accordingly, when the patient comes to the emergency clinic, the specialist can refresh the soundness of the cloud framework and effectively do the vital treatment process. Cost reserve funds, better administration, and opportune meticulousness are strong ways of further developing patient consideration principles.

3. Network safety in HIE and PHI Protecting medical care data is a critical worry in the medical care industry. Changing this data to the cloud can give the perfect proportion of safety to safeguarded wellbeing data (PHI) and medical services data trade (HIE) frameworks. Organizations giving distributed computing to medical services dissect and execute measures that solid cloud facilitates. In this way, distributed computing in medical care brings about information security and assurance from unapproved access.

Clinical Information Exchange (HIE) Sharing wellbeing data assists medical care associations with sharing the information that exists in their practically restrictive EHR frameworks. HIE can be sent utilizing a connection to an essential cloud execution. Distributed computing frameworks can be intended to make HIE safer than conventional client-server frameworks against the main sources of medical care information breaks. HIE framework in the change to distributed computing

- Versatile to the necessities of different divisions and the size of the organization.
- Decrease costly expenses.
- Compact for remote admittance to data and frameworks.

A likely application for Protected Health Information (PHI) distributed computing is to oversee admittance to individual wellbeing records (PHR) and electronic wellbeing records (EHR). This permits clients to get to the PHR data set through programming and offer PHR information. For instance, Microsoft HealthVault is a cloud-based stage utilized for PHR the board. The stage permits patients to store, make due, and share their PHRs by giving a helpful method for bringing wellbeing information from clinical gadgets into HealthVault without the utilization of go-between devices or programming. Clinical information can be handily observed, overseen, and got to through the UI. The program likewise has progressed sharing capacities that permit you to control one more degree of shared information, share different profiles, and speak with explicit clinical establishments, gadgets, or programming applications, all under the oversight of the information proprietor. Offers. This stage utilizes standard sharing conventions like SOAP, and CCR/CCD outfitted with developer interfaces so the wellbeing vault becomes helpful for work area and versatile applications.

4. Practical assets Cloud figuring in medical care offers better help in HR, authoritative and functional capabilities. It likewise guarantees the administrations with regards to booking, obtaining documents, references, and stock administration to make the interaction proficient, worthwhile, and savvy. Thusly, better asset distribution is conceivable for a minimal price utilizing cloud administrations.

5. Crisis adaptability Cloud administrations are very valuable in crises like pandemics, cataclysmic events, and long gridlocks. At such critical times, specialists and clinical considerations are deficient. Consequently, the data is shipped off the parental figure through the cloud framework, alongside directions on the best way to treat the harmed or wiped out.

6. Trailblazer of clinical innovative work Health suppliers, specialists, and medical care experts invest a lot of energy exploring to work on their training. They gather significant data about medical procedures and therapy. The putaway and secluded data serve to appropriately break down and explore them and make contextual investigations and papers for future reference to different doctors.

7. Production network upkeeps the huge measure of information connected with clinical gadgets, prescriptions and tablets should be appropriately made due. During capacity, the date ought to incorporate all sections, for example, lapse date, buy date, provider subtleties, and so on. You can save this whole information and recover it from the cloud. Cloud administration refreshes for gadgets and prescriptions that are going to lapse ought to be renewed right away.

8. Admittance to strong examination Cloud-based devices are helpful when the errand is scientific, as a significant piece of the examination comprises putting away and controlling the information. Well-being information, regardless of whether organized, is an immense product. Gathering significant patient information from various sources in the cloud is conceivable. Applying large information examination and man-made reasoning calculations to patient information put away in the cloud can help clinical exploration. High-level distributed computing permits huge datasets to be handled rapidly. Dissecting patient information can

likewise make ready for growing more customized treatment plans for patients. It likewise catches generally persistent information with the goal that it isn't neglected while recommending treatment.

9. Patient responsibility for Cloud registering concentrates the information and gives patients command over their wellbeing. It helps patients support in settling on conclusions about their wellbeing. As the reinforcements are computerized and there is no single touchpoint where the information is put away, recuperation of information becomes compelling.

10. Cooperative interoperability Moving clinical information to the cloud further develops joint effort, works with availability, and increments patient association. Distributed computing permits doctors and medical services suppliers to cooperate to work on persistent wellbeing. At the point when all information is free in one spot, cooperative units are shaped between partners to design, track progress and give reliable consideration to further develop wellbeing. Patient clinical history, analysis, and other clinical records are in a flash common with different clinical experts through the cloud. Working with a health care coverage organization can likewise prompt clinical offices that compensation without holes.

3.7 *How Does Cloud Computing Work in Healthcare?*

Distributed computing gives a steady framework to emergency clinics, clinical practices, inclusion organizations, and studies offices. The significant objective at the rear of upgrading registering sources at decline starter capital outlays is far. Additionally, distributed computing can diminish the limits to the development and modernization of medical services designs and applications. Over the long haul outcomes in making the overall wellness measurements control gadget more noteworthy bendy and versatile.

Contingent on the functionalities, some of the fundamental degrees of work process comprise of social occasion impacted individual's medical services measurements. The sensor hub on the impacted individual feature is chargeable for adapting to the overall insights of an impacted individual. This insight integrates the heartbeat, blood pressure, and physiological data of the impacted individual. The information is gathered through biometric hardware, which communicates it also to the wi-fi sensor hub. Subsequently, the insights get transferred to the cloud from a wi-fi sensor hub with the use of sensor measurements spread component.

The work process of the cloud contributions is presented from the mentality of individual and public cloud report situations. The confidential cloud stage comprises equipment and programming program added substances that are arranged with all perceived medical services prerequisites. The significant functionalities stressed in this work process are confirmation, approval, measurement tirelessness, insights trustworthiness, and measurement privacy.

- Confirmation
  It is performed through cryptographic conventions for allowing clients to get passage to individual cloud sources after a hit ID check.
- Approval
  This technique ensures the clients get passage to individual cloud sources.
- Information industriousness
  This trademark works to save the medical services insights for quite a while period premise.
- Information trustworthiness
  It ensures that the measurements are whole and steady at some stage in any change activity.
- Information privacy
  It outcomes in sending the measurements, which stays helpful best through the method of a method for the user. The following strides of cloud-based engineering give a total thought to the general work process.

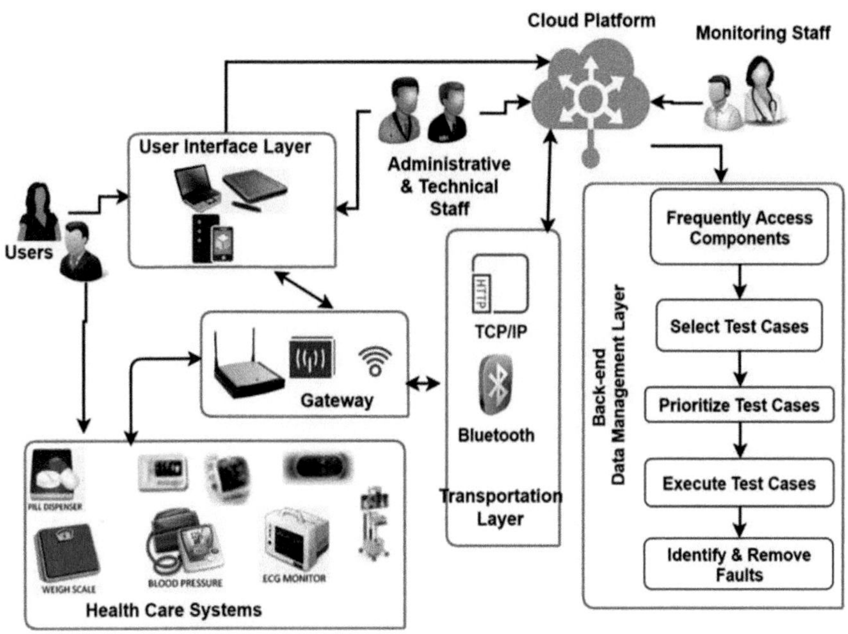

Fig. 4. Cloud computing architecture in Health Care.

Patient requests permission

Step 1: Just outside clients, like patients, and outsiders, for example, protection suppliers, drug stores, medical care research associations, and medication producers, can get to public cloud administrations. An extra outer client is a patient. To present a solicitation for authorization, he interfaces on with (username and secret phrase) tending to public personality and access control cloud administrations.

Step 2: The public cloud processes the solicitation before sending it to the confidential cloud organization.
Contingent upon the sort of solicitation for capacity, access, or handling of wellbeing information, it is taken care of at the public cloud level and sent from the confidential cloud to personality and access control administration.

Step 3: The solicitation is endorsed or denied. A medical care private cloud application server gets the solicitation if a confidential cloud server acknowledges it. Then again, on the off chance that the solicitation is declined, a warning message is given illustrating the justification for why.

Step 4: The specialist requests consent. The specialist is viewed as an inward client. Subsequently, he signs into the confidential cloud benefits and sends character and access control an approval demand that incorporates the client name and secret word.

Step 5: The cloud application server processes the specialist's solicitation to get the information.
Doctors can get the information from the public cloud application server once the validation is effective, so, all in all, confidential cloud administrations handle the solicitation.

Step 6: The patient is sent a clinical suggestion straightforwardly. A specialist might give the patient direct criticism as suggestions or solutions. It is critical to foster these sorts of cloud-based medical care frameworks for calamity readiness and provincial wellbeing. Moreover, medical services associations and clinical experts ought to begin embracing cloud-based advancements for clinical picture stockpiling and records on the board. The essential objective of this sort of arrangement is to assuage the clinical staff's weighty responsibility by giving better persistent consideration and clinical frameworks.

3.8 What are the applications of cloud computing in the healthcare industry?

Distributed computing displays a fundamental need to foster applications for superior execution of information handling and the executives. There

Fig. 5. Cloud computing services.

are various cloud administration contributions for medical services, covering a great many capacities. This pattern is featured through various applications and designs given the cloud in medical services. Following utilization of distributed computing in medical care is expected to drive a tech-drove medical services framework.

3.9 Cloud Computing Services

The use of the cloud has highlighted the urgent need for more comprehensive solutions for high-performance data processing and management. Many cloud provider services exist for the healthcare industry, with a wide range of capabilities. This approach is characterized by exceptional healthcare cloud-based products and architecture. The following healthcare cloud computing packages strive to mandate a tech-driven medical gadget.

1. Management Information Systems

The healthcare industry has started using control records structures to organize the flow of information both inside and outside of the organizations. The tool is used by doctors to deliver better patient care, by patients to inquire about the provider, by directors to manage the human resources, billing, and finances, and by top management to make forecasting and decision-making processes easier. Builders employ cloud-based platform services to expand, test and set up this gadget due to the secrecy of the data. It ensures quick team collaboration, cross-platform interoperability, and device connectivity with many legacy systems.

2. Telemedicine and Drug Discovery

These days, records and verbal trade innovation are blended to help and offer impacted individual consideration contributions. Distributed computing might be utilized as an ICT framework for telemedicine projects. Telemedicine innovations like telesurgery, sound/video conferencing, and teleradiology have made it critical to team up and talk among medical services partners. Telehealthcare contributions help the victims to get clinical cures in any area. Moreover, logical specialists can extend their expert assessment to adapt to confounded logical cases. Cloud-essentially based absolutely programming program grants doctor impacted individual and doctor exchange related to transmission and chronicling of logical pictures. Cloud-principally based thoroughly replies in telemedicine are important:

- To retrieve the archives at their location and time.
- To share patient scientific data in real-time across geographic boundaries.
- To reduce time- and money-wasting doctor appointments that are unnecessary.
- The drug development process needs adequate computational power to go through trillions of chemical structures and identify the compounds that can be made from them. Numerous cloud-based IaaS services significantly contribute to the simplification of this process. For instance, a collaboration between Molplex, Newcastle University, and Microsoft Research has used the IaaS cloud to speed up and reduce the cost of drug discovery.

3. Digital Libraries

- Libraries serve because of the inventory of understanding for logical understudies, analysts, and experts. The paper-essentially based logical libraries in developing countries can't meet the financial boundaries. Consequently, cloud-principally based virtual libraries give a major amount of record stockpiling, ordering supplier, question dialects, and website facilitating supplier library control structures. Cloud-fundamentally based answers are applicable to digitize libraries for:
- Organizations and individuals benefit from the capacity on request.
- Data searchers make the looking technique smooth by utilizing a semantic-basically based thoroughly question approach.
- Doctors to get conscious of present-day improvement withinside the logical region and subsequently upgrade their artistic creations practice.

4. Virtual Medical Universities

Due to its adaptability and pay-as-you-go business model, cloud computing has maintained its popularity during the current academic quarter. Numerous initiatives have been developed to support instructional learning both on and off campus. Medical colleges may employ cloud-based models to deliver online courses, conduct behavior seminars, and foster global academic cooperation. Clinical settings can achieve numerous first-year students at a low cost and with a lot less work using the aid of cloud computing.

5. Clinical choice guide device (CDSS)

An expert gadget follows a clinical expert's skill and leads to propose the impacted individual record examination. Doctors utilize this gadget for anticipation and cure purposes. Distributed computing age can extend those designs to direct impact individual considerations in sync with the necessity. The production of phone age with coordinated sensors to music the coronary pulse, circulatory strain, and diabetes makes those cloud structures green for an ongoing guess. In this way, the victims can rate their data with the gadget and get appropriate guidance. When followed with cloud contributions, those designs can make specific suitable cures way, especially in the event of a crisis, while clinical specialists aren't available [23]. For instance, Brayen progressed to a cloud-principally based CDSS at the records control and sharing system. It changed into intended for a storehouse exact through Clinical Decision Support Consortium (CDSC). This vault covered the data related to hypertension, diabetes, and coronary corridor infections and is essentially founded absolutely on an organization cloud facilitated through Partners HealthCare. This structure stressed the cloud website facilitating principally based motor, which saved a controlled data set of number one consideration victims. The fundamental explanation of this stage changed into supplying preventive consideration warnings to clinicians and surrendering clients.

6. Population Health control

Cloud contributions help to music ailments, map them geospatially and tell the general population wherein danger exists. Medical care organizations can authorize those contributions through the utilization of distributed computing. The gadget that exists withinside the market these days is Disease Control and Prevention (CDC). The CDC offices are progressed as a feature of a reconnaissance strategy to sell public wellness observation challenges [18].

7. Better Practice Management

Using the cloud, healthcare organizations may coordinate, manage, and improve their administrative and financial plans. It increases productivity by automating routine tasks to make the process simpler. Clinical professionals and care providers may get a ton of information using cloud services to create effective treatment regimens. Patients can make use of such services to identify their medical issues and communicate with their treatment provider. For instance, the Flatiron oncology cloud is a next-generation solution advanced for the value-based treatment of cancer patients [17].

8. Education in Health

The public may get information about the problems with hygiene, food, and fitness on the internet. The majority of consumers find trustworthy sources for health-related information online, such as websites, support groups, and blogs that are dedicated to certain illness types. Patients who have already coped with a condition might help new patients by sharing their knowledge, experience, diet, and pharmaceutical regimen as a kind of self-care. Cloud-based PaaS and SaaS applications may educate and train the general population about self-care. Hosting services are provided to these resource owners. The volunteers might also pay for cloud services like forums and chat rooms to start their club.

9. Biological Softwares

The issues with immense information in organic programming are settled as distributed computing draws near. The bioinformatics cloud is the term for this methodology. The information stockpiling, securing, examination, and enhancement of life science calculations as well as information serious logical instruments in bioinformatics are undeniably given by the cloud model made for natural programming.

In the end, the information is all over the place, and the medical services area is no exemption. There is huge potential for distributed computing in the medical services area as cloud innovation grows dramatically. A human's key right currently incorporates admittance to top-notch medical care. Moreover, this is an expensive and very involved topic [19]. Indeed, even the most evolved countries on the planet battle to meet the gigantic measure of medical care needs.

In any case, coordinating distributed computing influence into the medical care framework can without a doubt bring about genuine headways that benefit everybody's admittance to superior grade, modest medical care as opposed to only a little gathering of rich people.

4. IoT and Cloud Data Extraction

4.1 IoT—Primarily Based On E-Health Systems Architecture Elements

One of the essential plan constraints for IoT-based completely E-Health structures is the design [4]. As a rule, we order medical services design into classifications, for example, circulated and cloud-based totally models. Because of the circulated idea of EHRs, we never again give any records on a unified structure. It is spread across at least a couple of substances, such as emergency clinics, patients, doctors, and exploration experts, for different purposes. In IoT-based altogether E-Health frameworks, this delivers the utilization of unified structure repetitive.

4.1.1 Distributed Architecture

A decent number of servers are associated with each other and each has its interesting functionalities and style that make up the scattered construction. Since EHRs are very scattered in nature, this sort of design helps IoT-based completely E-Health structures. A gathering of PCs cooperates to achieve a typical point in a circulated structure. Clinical and sanatorium control structures are the most appropriate for this kind of organization [20]. The dispersed servers assemble and store the patient wellness records from the nearby clinics. The focal server then cycles and examinations the measurements that were saved at the appropriated servers for use in exploration and operations. This construction's disadvantage is that it is as of now not especially reasonable for ongoing wellness global positioning frameworks. Dispensed structure supports accelerating execution by and large and needs increasing methodologies. Furthermore, while managing an enormous volume of intricate insights, its general presentation endures. This is because of the way that it just backings level scaling and disregards vertical scaleup.

4.2 Cloud-Primarily Based Architecture

There is a dependence on cloud specialist co-ops or a focal expert in a framework that is subject to the cloud (CA). It uses half-breed, private, and public cloud frameworks. The edge region networks are utilized in cloud-based models to screen the patient's wellbeing data (BAN). Through the wearable savvy gadgets, the casing sensors are projected into and across the subject's body. The casing sensors track the patient's wellness information, which is then shipped off a cloud server for capacity purposes. Through cell phones, the patients can likewise change their wellness information. The health care coverage supplier firm arranges a Service Level Agreement

(SLA) with the patients and directs the administration of EHRs on a cloud server. There is a reliance on distributed computing in a cloud-based generally speaking structure. With regards to distributed computing, the cloud transporter organization (CSP) gives wellness care administrations. The distributed computing foundation is utilized by the CSP to oversee and look for EHR. After the information is saved money on the cloud server, the CSP applies rules for access and stores it with a dependable gathering of clients. Just when the CSP characterized admittance strategy lines up with the singular data access strategy is the individual ready to get to the data without any problem. For the most part, talking, there are selective methodologies as well as impacted individual driven and non-impacted individual driven ways. In a patient-focused approach, the patient determines the entrance rules and offers the data. While managing crises, this approach ends up being inadequate. The proposed gadget in this manner presents the CSP-focused approach, wherein the Cloud Service Provider (CSP) gives the data to gain admittance to strategies and takes care of the methodology engaged with the administration of wellbeing data. Moreover, the cloud-based generally speaking design keeps on being the most reasonable reaction for the IoT-based by and large E-Health structures since it is fit for taking care of huge and complex EHR in a harmless to the ecosystem way. Figure 6 assesses the security of an IoT-based E-Health System.

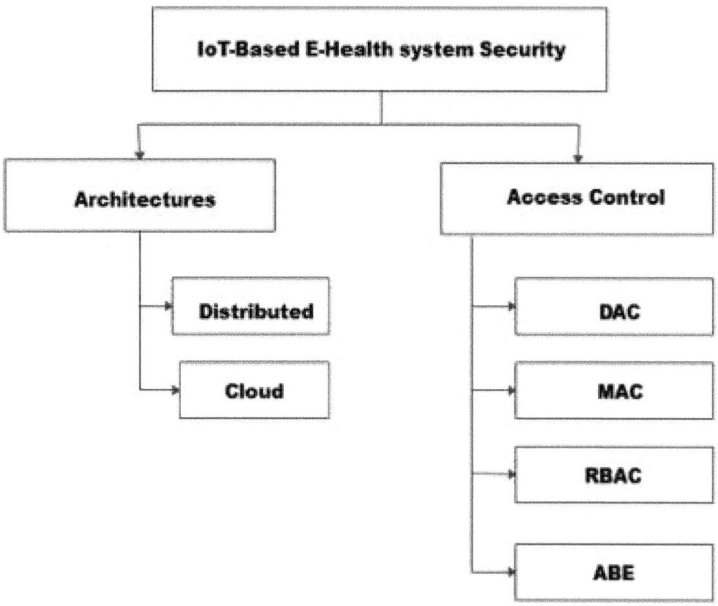

Fig. 6. Overview of IoT-based e-health system security.

5. Blockchains and IoT

For the steadily developing IoT device climate to be suitable, the creators contend in [6] that a push toward a decentralized construction is fundamental. From the stance of the producer, the ongoing concentrated rendition has a high upkeep cost – remember the dispersion of programming moves up to the huge number of gadgets for a long time after they were gradually gotten rid of. The requirement for a "wellbeing utilizing receptiveness" approach and a reasonable [7] absence of confidence in items that "telephone home" with inside the set of experiences are both felt by the client. The creators successfully bring up that a blockchain offers a popular answer for this issue and that these issues can be settled with a versatile, trustless distributed model that can work straightforwardly and spread data securely. To dive deeper into how this will work, examine the arrangement underneath. The single blockchain local area drives a maker's all's IoT gadgets. The maker utilizes a finesse plan that empowers them to sell the hash of the latest firmware and move up to the local area. The gadgets either come pre-modified with the brilliant understanding's location installed into their blockchain client or they find out about it through a revelation administration. Then, at that point, utilizing a disseminated distributed filesystem that incorporates IPFS, they might question the understanding, find out about the refreshed firmware, and solicitation it by its hash. The producer's hub could satisfy the underlying solicitations for this report (while likewise adding to the local area), yet after the double has spread to adequate hubs [10], the maker's hub can quit doing as such. A gadget that joins the local area after the maker has quit partaking in it can in any case recover the ideal firmware substitution and be certain that it isn't the right report, expecting the gadgets to be empowered to share the paired they got. With no human collaboration, everything happens consequently.

5.1 Architecture Concept

In IoT-based medical services structures, various compositional plans have been put out [5, 12, 13], albeit most of them aren't harmless to the ecosystem because of their security elements and data taking care of systems. As of late, a couple of structures [14, 15] that are centered explicitly around safeguarding sensor information and using a couple of cryptographic strategies have been added. In this review, we propose a unique strategy for getting any IoT-based gadget and its interior correspondence organization. In our machine, we propose engineering (Fig. 7) that is constructed completely and chiefly on a steady entryway that controls the whole gadget. Here, the Arduino MKR zero microcontroller

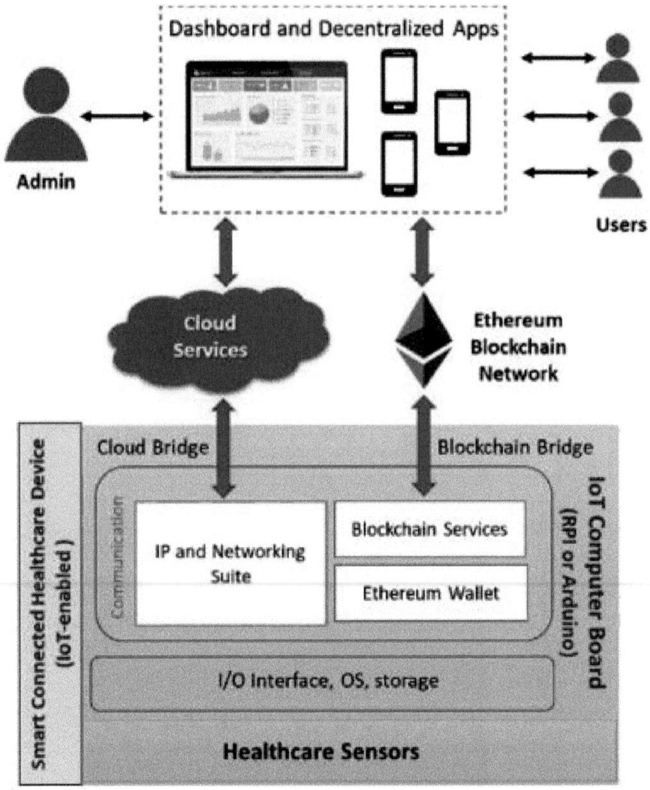

Fig. 7. Architecture of IoT based secures E-health system.

board is utilized with the steady entryway. We utilize temperature, heartbeat/heartbeat, muscle, and blood pressure sensors to assemble the casing boundaries of a competitor. A pre-shared key will be utilized by sensors to gather the information and speak with the entryway. During the device design process itself, this pre-shared key can be unique.

The gateway will then check the data it has gathered involving a versatile rule motor for any abnormalities. A clever key trade method that is completely founded on the LEACH convention was made conceivable by the capacity to rapidly and dependably inform the following device of any irregularities that were found. Entryway then sends the information to a distributed storage office for additional utilization. Using the changed HIP-DEX convention, the channel between the entryway and cloud is made conceivable. The wellness information of an individual partaking in a game can habitually be shown by end clients (specialists, physiotherapists, or some other lawful individual) utilizing an Android application. The utilization of the altered HIP-DEX convention by and gets correspondence between the end-client device and the cloud.

6. Conclusion

The blend of blockchains and IoT, as we've shown, can be fairly powerful. Blockchain gives powerful, really disseminated shared structures and the ability to associate with companions in an auditable, trustless way. We can robotize complex, multi-step processes thanks to savvy contracts. The IoT climate's resources with the actual world are the gadgets. At the point when they are consolidated, we might robotize difficult exercises in novel and imaginative ways, acquiring cryptographic undeniable nature as well as tremendous expense and time reserve funds during the cycle. We guess that the nonstop reception of blockchain innovation in the IoT space will bring about critical upgrades across numerous ventures, introducing new business drifts and constraining us to reevaluate how we presently apply designs and cycles. the making of new savvy assessments for the dependable treatment realities and inclusion charging realities units, and their quick circulation to the overall population. Our answer offers auditable e-Health information while keeping up with patient security and insurance. While government and administrative associations are allowed extra personalities for review and congruity purposes, clinical specialists are allowed to get to the huge e-Health blocks. The huge number of designers and the elevated degrees of interest inside the organization will ultimately drive the blockchain age into a well-ordinary method of activity in the e-Health space. Our answer is a decent exertion that likewise further develops execution and reliability, which fall into place easily from the idea of blockchains. More custom-made prescriptions are made conceivable by utilizing the total and forward-thinking information blocks accessible to all concerned assistance sellers on account of the computational rationales remembered for the e-Health blockchains.

References

[1] Md Ashraf Uddin, Andrew Stranieri, Iqbal Gondal and Venki Balasubramanian. 2021. A survey on the adoption of blockchain in IoT: Challenges and solutions Blockchain: Research and Applications, 2(2): 100006.

[2] Nguyen, D.C., Pathirana, P.N., Ding, M. et al. (2019). Integration of Blockchain and Cloud of Things: Architecture, Applications and Challenges arXiv.

[3] Ellouze, F., Fersi, G. and Jmaiel, M. Blockchain for the internet of medical things: A technical review.

[4] Jmaiel, M., Mokhtari, M., Abdulrazak, B. et al. (eds.). 2020. The Impact of Digital Technologies on Public Health in Developed and Developing Countries, 12157, Springer, Cham, France (2020), pp. 259–267.

[5] ur Rehman, M.H., Yaqoob, I., Salah, K. et al. 2019. The role of big data analytics in the industrial internet of things. Future Generate. Comput. Syst., 99: 247–259.
[6] Panda, S.S., Satapathy, U., Mohanta, B.K. et al. 2019. A blockchain-based decentralized authentication framework for resource-constrained IoT device 2019 10th International Conference on Computing, Communication and Networking Technologies; 6–8 Jul 2019; Kanpur, India, IEEE, Piscataway, NJ, USA (2019), pp. 1–6.
[7] Liu, W., Mundie, T., Krieger, U. and Zhu, S.S. 2017. Advanced Block-Chain Architecture for e-Health Systems, 2017 19th International Conference on E-health Networking, Application & Services.
[8] KonstantinosChristidis and MichealDevetsikiotis. Blockchains and Smart Contracts for the Internet of Things, Special Section on the Plethora of Research in the Internet of Things (IoT).
[9] An Advanced Wireless Sensor Network for Health Monitoring, Varone, G., Wood, A., Selavo, L., Cao, Q., Fang, L., Doan, T., He, Z., Stoleru, R., Lin, S. and Stankovic, J.A. Department of Computer Science, University of Virginia.
[10] Tamizharasi, G.S., ParveenSultanah, H. and Balamurugan, B. 2017. IoT-based E-health system security: A vision architecture elements and future directions. International Conference on Electronics, Communication and Aerospace Technology ICECA.
[11] Highly Secure and Efficient Architectural Model for IoT Based Health Care Systems Binu P. K., Karun Thomas, Nithin P. Varghese, @ 2017 IEEE.
[12] Brody, P. and Pureswaran, V. 2014. Device democracy: Saving the future of the Internet of Things. IBM Institute for Business Value, Tech. Rep., Sep. 2014. [Online].
[13] Angwin, J. 2015. Own a Vizio Smart TV? It's Watching You.
[14] Benet, J. 2016. IPFS—Content Addressed, Versioned, P2P File System (DRAFT3), accessed on Mar. 15, 2016.
[15] IPFS—Content Addressed, Versioned, P2P File System, accessed on Mar. 15, 2016.
[16] Benet, J. (2015). Replication IPFS — OrtheBackingUpContentModel.
[17] Nikolaevisjiy, D. Korzun and Gurtov, A. 2014. Security for Medical Sensor Networks in Mobile Health Systems. IEEE Int. Symposium on a World of Wireless, Mobile and Multimedia Networks, 2014.
[18] Rasid, M.F.A. et al. 2014. Embedded gateway services for Internet of Things applications in ubiquitous healthcare. 2014 2nd Int. Conf. Commun Technol. ICoICT, pp. 145–148.
[19] Kiran, M.P.R.S., Rajalakshmi, P., Bharadwaj, K. and Acharyya, A. 2014. Adaptive rule engine based IoT enable remote health care data acquisition and smart transmission system. 2014 IEEE World Forum Internet Things, WF-IoT 2014, pp. 253–258.
[20] Sriram Sankaran. 2016. Lightweight Security Framework for IoTs using Identity-based Cryptography. 2016 Int. Conf. on Advance in Computing, Communications, and Informatics (ICACCI), 2016, pp. 880–886.
[21] Sriram Sankaran and Ramalingam Sridhar. 2015. Modeling and analysis of routing in IoT networks. 2015 Intl. Conf. on Computing and Network Communications (CoCoNet'15), pp. 649–655.
[22] Lakshmi Mohan, Jinesh, M.K., Bipin, K., Harikrishnan, P. and ShijuSathyadevan. Implementation of Scatternet in an Intelligent IoT Gateway. Adv. Int. System and Compt.

Chapter 7

Machine Learning Model for Predicting Viral Diseases using Linear Regression

R Gomathi

The outbreak of new viral diseases like COVID-19 become common in our country right from 2019. New viruses and its mutated varieties are increasing day by day which in turn affects the universal economy and the well-being of people all over the country. Statistics show that the death rate is increasing due to this viral spread. A timely prediction of such kind of viruses and their precautionary measures has become more important in today's life. This research projects about prediction model which depends on machine learning for the timely identification of viral infections in patient's based on some data provided by the diagnosis of the patients. The examination is performed for a specific kind of virus called Corona virus. The study was performed using the COVID-19 datasets and a linear regression algorithm was planned to guess the total count of people affected with this disease. Initially, the data were pre-processed using several pre-processing techniques. Also, some validation techniques are applied in order to guess the disease accurately. Experiments were performed using COVID-19 affected people data. The consequences disclosed that

Associate Professor, Department of CSE, Bannari Amman Institute of Technology, Sathyamangalam.
Email: gomsbk@gmail.com

the model predicts very accurately and the metrics are recorded. This research helps the medical professionals to make decisions related to the treatment of the disease and safeguard the patients from the effect of modern viruses.

1. Introduction

Viral diseases mostly occurs due to the blowout of various pathogens from human to human, animals to animals and from animals to human beings. There are abundant different ways in which such kind of transmission occurs. When we contemplate the speed of transmission, it is actually fast. It is difficult to diagnose such viral diseases at the early juncture and prevention and control is another major challenge in this problem.

Corona virus [1] is one such virus which causes the Corona virus disease-COVID-19. The outbreak of this disease had a high influence on the economy of many countries including India.

In continuation to the blowout of infectious diseases all over the sphere, some kind of scientific analysis is required to predict such kind of viral spread and prevention of those diseases. Machine learning (ML) algorithms proposed in research tends to predict the extent of virus based on the weather conditions and its related parameters [2]. From the analysis it was found that with a high value in temperature, the figure of viral infected count becomes low.

The symptoms of this sickness include coughing, temperature, weakness, littleness of breath, loss of taste and smell, and frequently no indications. The condition principally affects the inferior and upper respiratory systems. The lungs are the main organs impacted by the illness. Rendering to the MERS (Middle East respiratory syndrome) and SARS (Severe Acute Respiratory Syndrome) incubation periods, the infected person exhibits symptoms between 2 and 14 days. This puts patients at a significant risk of passing away.

In China, the coronavirus (COVID-19) first appeared in December 2019. Over 95 million cases had been found globally as of January 2021, with a fatality rate of 12% of all closed cases [8]. The pandemic's quick spread is a major concern for the entire world and poses a serious danger to both community health and the global economy. Most nations limited social collaboration through preventative measures like segregation and quarantine to stop the sickness from spreading. However, because of late detection and the unusual and uncharted behavior of the disease, many diseased people did not get anything from the appropriate treatment. Recently, several researchers concentrated on creating novel screening techniques for infected patients at various phases in order to identify noticeable correlations between the patient's clinical characteristics and the likelihood of dying.

According to recent research investigations, Machine Learning (ML) and Artificial Intelligence (AI) algorithms can be crucial in limiting the influence of virus dissemination. There are many various research trajectories that ML solicitation technologies on patient data may be categorized under [9]. Predicting contamination and mortality rates and developing a prototype to categorize patients based on their medical findings are the most significant research directions [10, 11]. These investigation studies are crucial and will help those working in the healthcare industry be well prepared and yield all essential safeguards to stop the plague from spreading.

The scientific characteristics of patients were investigated in research using ML tools. Some key factors were used to envisage the impermanence rate of individuals [3]. Real time datasets were taken and tested using the machine learning models.

2. Related Work

Early discovery and judgment using machine learning techniques helps us to avert the wide spread of viral diseases and to fight the pandemic with diverse methods of collecting data like X-ray, experimental, CT scans, and plasma mockup data.

The critical situation and likelihoods of patients survival with simple COVID-19 contamination was predicted depending on danger factors and geographic evidence. The research used the dataset which contains evidences from patients admitted in the Tongji Hospital since January to February, 2020, which includes 201 survivors and 174 expired within the same period. The research used an XGBoost (XGB) algorithm [4] to identify three main clinical features. The proposed model was experimented with data from 29 patients and it was found that the model's ability to guess the risk of death was found with 0.94 precision and 0.91 prediction accuracy. These kinds of models help physicians with a device for identifying dangerous conditions, and thereby tries to reduce the transience rate.

Deep learning (DL) and machine learning (ML) are increasingly being used in the arena of medical science, notably in the areas of audio, video and linguistic data [14]. By combining a DNN model combined with two machine learning models for virus prediction using test center results, there was a hope to generate a new, improved ensemble model. Based on worth counts, medical importance-related structures, and absent values, 86 characteristics were chosen from datasets. With 5145 cases, sample datasets, which included 326,686 test center results were gathered. With respect to the Worldwide grouping of Diseases, 10th Modification (ICD-10) codes, a total of 39 distinct diseases were examined.

The XGB ideal was used with some additional dataset to envisage the severe and the death cases and also to classify the risk features of COVID-19. The dataset was taken from United Kingdom Bio bank (UKBB) which includes 93 different variables collected between March and July 2020. Two types of studies have been undergone [5] based on the sample's groups. In the first type of study, the data consists of clinical prediagnostic data of 1747 COVID-19 affected patient records with both severe and death cases. The accuracy achieved for the severity class was 0.668, and for the fatality class, it was 0.712. In the next type of study, the data consists of the undesirable cases. The same model was applied, and the accuracy achieved was 0.669 for the severity class and 0.749 for the fatality class, respectively. The researchers tried to identify the five greatest important risk factors for severe cases and death cases, with age being the top factor for both cases. The other factors in this research include obesity, multiple comorbidities, impaired renal function, and cardio metabolic abnormalities.

SARS-CoV-2, also known as the Corona Virus, has wreaked devastation all over the world and things are just getting poorer. Its a pandemic illness that is daily dispersal from person to person. As a result, it's critical to monitor the quantity of impacted patients. The current approach provides computerized data in a collective manner that makes it exceedingly challenging to assess and forecast the blowout of disease in a specific location and globally. To report this issue, machine knowledge techniques were employed to effectively map the disease and its course. By examining patients' chest X-ray images, machine learning, a subfield of computers, is essential in correctly identifying those with the illness [13].

A prediction model was developed [6] using the support vector machine (SVM) in research to predict the simple problems of COVID-19 patients. In that study, the clinical and laboratory features were used. Using a few cases of COVID-19 patients, the main features were predicted to differentiate between the mild and severe cases. It was observed that the proposed model predicts patients with an accuracy of 0.775.

Another research was performed in research which implemented the SVM model [7] to classify the COVID-19 patients based on the harshness of the indications. The research applied SVM for the binary class label on a total of 130 records which includes urine and blood test results. It was a combination of both ill patients and patients with mild symptoms. The results proved that around 32 factors had high connections with severe COVID-19, and the accuracy was about 0.815. It was also originated in research that among all the factors like age and gender mostly affects the classification of cases as plain or minor. The research also predicted that patients aged around 65 were found with more severe cases than others.

And also it was found that the male patients have a sophisticated danger of developing severe COVID-19 symptoms. When the urine and blood test samples were considered, blood test result skins showed substantial differences between plain and mild cases.

The new coronavirus (COVID-19) epidemic had a severe effect on both the health of entire communities and the world economy. Despite the high COVID-19 survival rate, there are more severe instances that end in death every day [12]. It is anticipated that early identification of COVID-19 at-risk patients and the implementation of preventative interventions will improve patient survival and lower fatality rates. A study was performed which offers a strategy for predicting COVID-19 patient outcomes early on the foundation of data from home-monitored patient characteristics collected during quarantine. 287 COVID-19 samples of patients from the King Fahad University Hospital in Saudi Arabia were made use in that study. The classification techniques, namely logistic regression (LR), random forest (RF), and extreme gradient boosting, were used to examine the data (XGB).

Using artificial intelligence algorithms, 200 quantitative CT structures of COVID-19 pneumonia were compiled [15]. The COVID-19 guidelines were used to define the seriously ill cases. The predictors of serious illness were chosen from the clinical and radiological features using the least absolute shrinkage and selection operator (LASSO) logistic regression, respectively. So, using the following machine learning classifiers, clinical and radiological models: naive bayes (NB), linear regression (LR), random forest (RF), extreme gradient boosting (XGBoost), adaptive boosting (AdaBoost), K-Nearest Neighbour (KNN), kernel Support Vector Machine (k-SVM), and Back Propagation Neural Networks (BPNN) were created. Using the eight classifiers indicated above, a composite model including the chosen clinical and radiological parameters was also created.

2.1 *Prediction in Machine Learning*

When predicting the likelihood of a explicit conclusion, machine learning refers to prediction as the result of an procedure that has been followed on past data and applied to current data. The pace at which data is processed and evaluated is accelerated by machine learning. With very slight deployment adjustments, predictive analytics algorithms can now train on even larger data sets and do more in-depth research on a variety of aspects. If classification involves categorizing data, prediction is fitting a shape to the data that is as near to the definite data as possible. In its place of a fence that divides two bodies of data, the object we're fitting is extra like a skeleton that runs through one body of data. As before, the question of What? is provided by the algorithm, and the question of Where? is delivered by the data. The predicted value is read off the line as soon as new data points are acknowledged and inserted into the recipe.

2.2 *Regression for Predicting Viral Diseases*

Regression is a method of predictive modeling technique in which a significant association between a dependent mutable and one or more independent variables were found out. The various types of regression techniques available are Linear, Logistic, Polynomial, Ridge, Lasso, and Softmax. Linear regression is a direct model, in which this model adopts a lined affiliation between the input variables (x) and the single output variable (y). It means that y can be premeditated from a linear combination of the input variables (x). In simple linear regression model, once we have a single input, which uses statistics to estimate the coefficients. For this we will first calculate algebraic properties from the data like mean, standard deviation, correlation and covariance. One or more self-governing variables called predictor variables and one or more dependent variables called criterion are related in regression analysis, a statistical method. A predicted value for the criterion is obtained from the analysis as a outcome of a linear combination of the predictors.

Outcomes can be predicted using regression equations. Regression equations are essential to the statistical output following model fitting. The coefficients of each independent variable in the equation describe its relationship to the dependent variable. You can additionally factor values for the independent variables into the equation to forecast the mean value of the dependent variable.

Linear regression is unique of the most widely used and simple Machine Learning techniques. It is equally a statistical strategy and a method of predictive analysis. As implied by its name, the linear regression algorithm displays a linear connection between a dependent variable and one or more independent variables. The variation in the dependent variable's value as a purpose of the independent variable is discovered via linear regression since it indicates a linear relationship.

The general method for using regression to make decent predictions are:

- To build on the work of others, do some research on the topic. The ensuing steps are made easier by this study.
- Gather information about the pertinent factors.
- Please describe and evaluate the regression model.
- Use a ideal that appropriately turns the data to generate predictions if there is one.

While the method needs more effort than the intellectual method, it offers noteworthy advantages. Regression allows us to evaluate the bias and accuracy of our predictions:

- A algebraic model with bias makes forecasts that are consistently either too great or too short.
- The precision measures how well the forecasts match the definite numbers.

The objective while using regression to make predictions is to bring fallouts that are both normally accurate and near to the actual values. Data are available. Checking for a statistically significant association between the variables is the next step. If we wish to forecast the rate of another variable using the value of one variable, relationships or correlations between the variables are essential. The regression model's suitability for prediction needs to be assessed as well.

2.3 *Types of Linear Regression*

2.3.1 Simple Linear Regression Model

With just one idea, we can practice measurements to estimate the coefficients of simple linear regression. We must perform the necessary statistical analysis on the data to determine means, standard deviations, correlations, and covariance. To traverse the statistics and do statistical calculations, it must all be accessible.

2.3.2 Ordinary Least Squares

This Ordinary Least Squares can be used to estimate the constant values when we have several inputs. The aim of this method is to reduce the total squared residuals. Assuming a regression line crosses the data, it says to square the distance between each data point and the line and then add the squared errors for all the data points. This amount is attempted to be minimised via ordinary least squares. This approach treats the data as a matrix, and linear algebraic operations are used to estimate the values of the ideal quantity. You need access to all of the data and enough RAM to be able to fit the data and perform matrix operations.

2.3.3 Gradient Descent

If one or more inputs are available, you can use an approach to iteratively reduce the model's error on your preparation data to optimise the coefficient values. The process, known as Gradient Descent, works by beginning with arbitrary values for each coefficient. The sum of the squared errors is computed for each set of input and output values. The learning rate is the scale factor that is used, and the coefficients are updated to reduce the inaccuracy. The technique is repeated until a minimum sum squared error is reached or no more improvement is possible.

The magnitude of the upgrading step to be taken on each repetition of the procedure must be determined by a learning rate (alpha) constraint when employing this method. Because it is relatively simple to comprehend, linear regression models are recurrently used to teach gradient descent. In actual use, it is helpful when you have a dataset with a very great number of rows or columns that would not fit in memory.

2.3.4 Regularization

Regularization algorithms are modifications of the direct model's preparation. These algorithms purpose is to lower the difficulty of the model while also minimizing the sum of the squared errors of the classical on the training data (using conventional least squares).

2.4 Preparing Data Suitable for Linear Regression

There is a wealth of works on how your data should be organized to brand the greatest use of the model since linear regression has been explored in great detail. As a result, discussing these demands and expectations involves a high level of knowledge, which is scary. When employing Ordinary Least Squares Regression, is the most popular application of linear regression, in exercise, you can use these guidelines more as general guidelines.

The following are some of the approaches of preparation.

Assumption of linearity: The assumption behind linear regression is that there is a linear connection between the input and the result. It doesn't maintain anything else. Though it can seem apparent, it is a useful thing to remember when you have a lot of positive traits. The data may need to be changed in order to brand the relationship linearly. The log transmute for an exponential connection, for instance.

Eliminate noise: The assumption behind linear regression is that the input and output variables are both clean. Utilize data cleansing procedures that enable you to more clearly reveal and define the signal in your data. This is crucial for the output variable, and if at all possible, you should get rid of outliers there.

Rescale your inputs: Standardization or normalization strength help you get more accurate results from your linear regression.

Collinearity must be removed: When your input variables are extremely correlated, your results will be over fit using linear regression. For any contribution data, we can think about computing pairwise correlations and eliminating the most associated.

Gaussian Dispersion: If the input and output variables are distributed according to a Gaussian distribution, linear regression will yield more accurate predictions. You might gain something by applying transforms to your variables to give them a more Gaussian-looking distribution.

2.5 *Significant Assumptions for Linear Regression*

For linear regression inspection to be successful, the following assumptions must be taken into account: Consider each variable's mean, standard deviation, and number of applicable cases. For each model, it is important to analyse the regression coefficients, correlation matrices, partial and full correlations, multiple R, adjusted R2, change in R2, standard error of the estimate, analysis of variance table, projected values, and residuals. The variance-covariance matrix, the variance inflation factor, tolerance, the Durbin-Watson test, distance measures, the DfBeta and DfFit prediction intervals, as well as case-specific diagnostic data should all be taken into account.

When we think of scatterplots, partial plots, histograms, and normal probability plots when describing plots. Data should be quantifiable for both dependent and independent variables. The recoding of categorical variables, such as religion, primary field of study, or region of residence, into binary variables or other contrast variables is necessary. The distribution of the dependent variable need to be normal for each value of the independent variable. For all possible independent variable values, the variance of the distribution of the dependent variable should remain constant. All observations should be independent, and there should be a linear relationship between the dependent variable and each independent variable.

2.6 *Predictions with Linear Regression*

Predictive modelling, or more particularly machine learning, is primarily concerned with reducing a model's error or creating the most accurate forecasts possible at the expense of explainability. Applied machine learning will borrow, reuse, and even give away algorithms from several fields, including statistics, in order to accomplish these objectives.

As a result, machine learning has adapted linear regression, which was created in the ground of statistics and is examined as a prototypical for understanding the relationship between input and output numerical variables. It is a machine learning algorithm as well as a statistical algorithm.

In Linear regression, a variable's value can be predicted using linear regression analysis based on the value of another variable. The reliant on variable is the one you need to be able to predict. The independent variable is the one you're using to brand a prediction about the value of the other variable. With the help of one or more independent variables that can most accurately predict the value of the dependent variable, this type of analysis calculates the coefficients of the linear equation. The linear regression method minimises the discrepancies between expected and actual output values by fitting a line or surface. Simple linear regression calculators employing the "least squares" method can find the best-fit line given a collection of paired data. You then estimate the value of X (the dependent variable) using Y (independent variable).

The mathematical technique used in linear-regression models is straightforward and used to make predictions. Numerous corporate and academic disciplines can profit from the use of linear regression. In many fields, including business and the social, interactive, environmental, and biological sciences, linear regression is used. With the use of linear-regression models, future predictions may now be produced scientifically and with a high degree of dependability. Since linear regression is a statistical technique that has been around for a very long time, its features are well understood and can be learned fairly quickly.

Leaders in businesses and organisations can make better decisions by using linear regression techniques. Organizations collect a lot of data, and linear regression enables them to use that data to manage reality more effectively rather than relying on experience and intuition. Massive amounts of unprocessed data can be converted into knowledge. You can also utilise linear regression to give deeper insights by exposing patterns and connections that your coworkers may have previously noticed and taken for granted. For instance, studying sales and purchase information can reveal unique tendencies in buying on certain days or at certain times. Business executives can estimate times when there will be a spike in demand for their products using insights from regression analysis.

Assume that given the depiction of a linear calculation, for a specific set of inputs making predictions is as modest as solving the equation.

Consider an example in which we imagine that we are working to predict weight (y) from height (x). The linear regression model depiction for this type of problem would be like the following equation:

$y = A0 + A1 * x1$

or

Weight = A0 +A1 * height

in which A0 is the bias coefficient.

A1 is the coefficient for the height column.

A learning technique is used to find a noble set of coefficient values. Once the coefficient values are found, we can pad in different height values to calculate the weight.

This method is not only a simple machine learning algorithm, but it plays a major role in statistics. It comes from the supervised learning, in which all input is associated with a target label. The task of the linear regression model is to fundamentally recognize the pattern and discover the best fit line that covers each (input, target) pair.

2.7 *Dataset Used*

The SARS-CoV-2 virus is the infectious disease known as coronavirus disease (COVID-19). The common of virus-infected individuals will experience a mild to severe respiratory disease and will recover without the need for special care. However, some people will get severe ailments and require medical attention. Elderly people and people with underlying medical conditions including cancer, diabetes, cardiovascular disease, or chronic respiratory problems are more likely to experience serious illness. COVID-19 has the potential to make anyone terribly ill or cause their death at any age. The greatest way to prevent or slow down transmission is to educate yourself on the condition and how the virus spreads.

By keeping a distance of at least one meter between people, donning a mask that fits properly, and often washing your hands or using an alcohol-based rub, you can prevent infection in both yourself and other people. Get your immunization when it's your turn, and follow any local instructions. The virus can spread from the lips or nose in tiny liquid particles when a sick person speaks, sneezes, sings, or breathes. From larger respiratory droplets to tiny aerosols, these particles are diverse. It's essential to limit yourself to your house and relax until you feel better while using good respiratory technique, such as coughing into a flexed elbow. The dataset applied for this research includes the international data collected from the WHO COVID-19 condition report and the statistics was also updated from the website covid19india.org. The dataset was collected in a CSV file format and uploaded in Colab notebook and implemented with the Python 3.8.2 software.

Input to the algorithm:

- Entire number of infected cases
- Entire number of active cases
- Recovery numbers

Output after prediction: Entire deaths and case fatality rates (CFR).

In demand to get a good predictive value data was taken for the top fifteen infected countries.

3. Experimental Results

The following analysis shows the death prediction for week five in the dataset. A modest regression model was used to predict the count of number of deaths from the input variable. The model was strong with $r = 0.84$, $R2 = 0.72$ & adjusted $R2 = 0.77$, $P < 0.001$, 95% CI: 1.31–2.60. Based on the max limit & the min limit of the confidence intervals, the minimum, maximum and average death counts for week five was computed. And so the week five death counts for our country India was predicted based on the available input data from the top fifteen infected countries.

The following table shows the prediction results of maximum, minimum and average predicted death counts for week six based on the equation of the linear regression model.

With 95% of Confidence Interval	Intercept value and Co-efficient value		Predicted Death count for Week five
Point of estimation (Mean value)	b0	191.633	208
	b1	1.857	
Point of estimation (Lower value)	b0	−228.32	−214
	b1	1.302	
Point of estimation (Upper value)	b0	610.54	637
	b1	2.502	

Using the predicted data of week five death count, it is used as input with a simple linear regression model to predict the week six death count outcomes. The model was strong with $r = 0.94$, $R2 = 0.92$, adjusted $R2 = 0.92$ and statistically important with P-value < 0.001, 96% CI: 1.12–1.55.

4. Conclusion

One of the administered machine learning techniques used to predict continuous values is the regression model. The definitive aim of the regression algorithm is to design a best-fit line or a arch between the data. Three topmost metrics are used for evaluating the trained regression model which includes variance, bias and error. In this research work, a simple linear regression model was applied to predict the demise count of people who are exaggerated with viral diseases especially Covid-19. The experimental results show the sturdiness of the algorithm in envisaging the death count for varying amount of data.

References

[1] Hao, Y., Xu, T., Hu, H., Wang, P. and Bai, Y. 2020. Prediction and analysis of corona virus disease 2019. PloS One, 15(10): e0239960.

[2] Malki, Z., Atlam, E.S., Hassanien, A.E., Dagnew, G., Elhosseini, M.A. and Gad, I. 2020. Association between weather data and COVID-19 pandemic predicting mortality rate: Machine learning approaches. Chaos, Solitons & Fractals, 138: 110137.

[3] Chowdhury, M.E., Rahman, T., Khandakar, A., Al-Madeed, S., Zughaier, S.M., Hassen, H. and Islam, M.T. 2021. An early warning tool for predicting mortality risk of COVID-19 patients using machine learning. Cognitive Computation, 1–16.

[4] Yan, L., Zhang, H.T., Xiao, Y., Wang, M., Sun, C., Liang, J. and Yuan, Y. 2020. Prediction of criticality in patients with severe Covid-19 infection using three clinical features: A machine learning-based prognostic model with clinical data in Wuhan. MedRxiv, 27: 2020.

[5] Kenneth, C.Y., Xiang, Y. and So, H.C. 2021. Uncovering clinical risk factors and prediction of severe COVID-19: A machine learning approach based on UK Biobank data. MedRxiv, 2020–09.

[6] Sun, L., Song, F., Shi, N., Liu, F., Li, S., Li, P. and Shi, Y. 2020. Combination of four clinical indicators predicts the severe/critical symptom of patients infected COVID-19. Journal of Clinical Virology, 128: 104431.

[7] Yao, H., Zhang, N., Zhang, R., Duan, M., Xie, T., Pan, J. and Wang, G. (2020). Severity detection for the coronavirus disease 2019 (COVID-19) patients using a machine learning model based on the blood and urine tests. Frontiers in Cell and Developmental Biology, 683.

[8] Worldometers-COVID-19 Coronavirus Pandemic. https://www.worldometers.info/coronavirus/?utm_campaign=homeAdvegas1? (Accessed January 17, 2020).

[9] Lalmuanawma, S., Hussain, J. and Chhakchhuak, L. 2020. Applications of machine learning and artificial intelligence for Covid-19 (SARS-CoV-2) pandemic: A review. Chaos, Solitons & Fractals, 139: 110059.

[10] Chowdhury, M.E.H., Rahman, T., Khandakar, A. et al. 2020. An early warning tool for predicting mortality risk of COVID-19 patients using machine learning. http://arxiv.org/abs/2007.15559.

[11] Nemati, M., Ansary, J. and Nemati, N. 2020. Machine-learning approaches in COVID-19 survival analysis and discharge-time likelihood prediction using clinical data. Patterns, 1(5): 100074.

[12] Aljameel, S.S., Khan, I.U., Aslam, N., Aljabri, M. and Alsulmi, E.S. 2021. Machine learning-based model to predict the disease severity and outcome in COVID-19 patients. Scientific Programming, 2021.

[13] Muhammad, L.J., Algehyne, E.A., Usman, S.S., Ahmad, A., Chakraborty, C. and Mohammed, I.A. 2021. Supervised machine learning models for prediction of COVID-19 infection using epidemiology dataset. SN Computer Science, 2(1): 1–13.

[14] Park, D.J., Park, M.W., Lee, H., Kim, Y.J., Kim, Y. and Park, Y.H. 2021. Development of machine learning model for diagnostic disease prediction based on laboratory tests. Scientific Reports, 11(1): 1–11.

[15] Liu, Q., Pang, B., Li, H., Zhang, B., Liu, Y., Lai, L. and Zeng, Q. 2021. Machine learning models for predicting critical illness risk in hospitalized patients with COVID-19 pneumonia. Journal of Thoracic Disease, 13(2): 1215.

Chapter 8

Patient Wellness: Ethereum Based Electronic Health Record Storage

CM Naga Sudha[1,*] and *J Jesu Vedha Nayahi*[2]

Medical and health related information about patients qualifies as sensitive and personal information. There are many laws such as HIPAA that prohibit sharing of a patient's intimate medical data to third party organizations without the full consent of the patient. With this in consideration, when the actual medical health record systems are analysed, vulnerabilities that threatens the confidentiality and integrity of the data were found. The other consideration about the current system is the fact that patients do not have control over their data. Therefore, a decentralized ledger can resolve such security issues, where patients can have control on their own data. Such a decentralization can be implemented by utilizing blockchain technology. Also, within a decentralized system, many number of participants would be connected as network participants. Hence, a permissioned blockchain system using Ethereum helps in facing these attacks. The proposed Ethereum based Electronic Health Record storage system uses InterPlanetary File System (IPFS), for storing the patient's sensitive

[1] Anna University-MIT Campus, Chennai, India.
[2] Anna University, Regional Campus, Tirunelveli, India.
Email: vedhaj2000@gmail.com
* Corresponding author: cmsudha30@gmail.com

data that can be encrypted and only after duly authorized by the patient, a person can view or change it. Further, the proposed Ethereum based Electronic Health Record storage system can be used to protect other kinds of sensitive data without being limited to health data.

1. Introduction

Blockchain is defined as a public ledger which consists of transactions that are shared among the network participants. A block present in a chain consists of block header and body. Block header constitutes version information, nonce, parent block hash, Merkle tree root hash and timestamp as shown in Fig. 1. Merkle tree root hash stores the hash of all transactions. Nonce is a value that gets rehashed according to the difficulty restrictions. It gets updated every time a block is added to the blockchain. The content of a block holds a transaction counter and the transactions. Blockchain is usually a peer to peer network and the records are immutable with a secure design. Adding new blocks or transaction records to a blockchain is called mining. Miners are the nodes that collect transactions and organize them into blocks in a network. Transactions are verified by the validators in blockchain. Blockchain uses public key cryptography which has a public and a private key to access the block records and assets. Smart contract is a program whose main purpose is to avoid the use of a third party between two entities. Blockchain based smart contracts are executed automatically when the given conditions are satisfied.

As blockchain is a decentralized system, consensus algorithms are used to ensure consistency of data across all peers in the network. Some of the commonly used consensus algorithms are Proof of Work (PoW), Proof of Stake (PoS), Practical Byzantine Fault Tolerance (PBFT) and Ripple. All these algorithms aim at selecting a trustworthy node from the network for publishing the transactions. PoW is based on the idea that a node which does heavy computation is less likely to attack the network. Nodes that are willing to participate in consensus need to solve a mathematically complex problem. A node which arrives at a value that is close to the nonce would be allowed to broadcast its transaction data. PoS follows a similar approach in which the nodes are required to prove their ownership with the help of currency. The cost of mining is greatly reduced in PoS when compared with PoW.

PBFT is used when every node is aware of every other node that is present in the network. PBFT is composed of three phases in which a node proceeds to the next phase only if it receives 2/3 of the total votes from other nodes. In Ripple, all nodes in the network are divided into

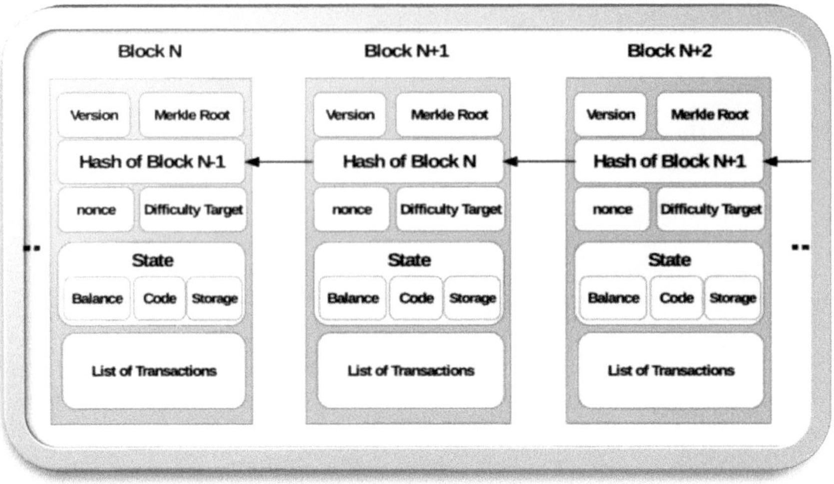

Fig. 1. Structure of a block.

servers and clients. Each server maintains a Unique Node List (UNL) that is queried every time before mining. The transactions are added to the ledger if they get approved by 80% of the nodes present in UNL. Blockchain plays an important role in healthcare and can help in improving medical applications, storing electronic records and monitoring devices. Healthcare providers and payers use blockchain to manage data and medical records which provide personal privacy and compliance. With a private blockchain, individual identity of patients and their sensitive medical records can be protected.

Bitcoin cryptocurrency came into existence in 2009 and as the code is open source, programmers are able to edit and improvise them. Blockchain technology evolution has different phases as follows:

- Blockchain 1.0:

  Distibuted Ledger Technology (DLT) contributed as first and noticeable use of blockchain technology where cryptocurrencies are focused more. Cryptocurrencies use transparent mechanism to monitor the technology.

- Blockchain 2.0:

  Digital Transactions are legally binded through policies called Smart Contracts. These are small computer programs that are integrated along with the Blockchain 2.0. Ethereum is the most prominent one.

- Blockchain 3.0:

  Decentralised Applocations are more focused on Blockchain 3.0 which avoids the centralized infrastructure.It stores and communicates via decentralised server. It popularises blockchain in various conventional sectors, health as well as education.

- Blockchain 4.0:
  Blockchain 4.0 provides solutions for the industrial demands of Industry 4.0 which involves automation, integration of various programs.

2. Related Works

Electronic Health Records (EHR) consolidates all types of medical reports that are stored in different formats such as JPEG, PDF which serves as data sources in different medical organizations. These can also include health information which are collected from wearable devices. Thus, EHRs are

Fig. 2. Taxonomy of Electronic Health Records (EHR).

defined as real-time records that are available for authorised users. EHR consists of various range of data that are used for diagnosis, demographic history, immunity level of patients, weight and medication history. Also, EHR whenever stored, it must define confidentiality, integrity and availability and must be accessible only when authorisation is provided. EHR storage system can reduce the loss of data but the security attacks are highly critical. Cyber-attacks are increasing day-by-day where the malicious users can target patient's wearable devices and inject any program to acquire control. EHR helps in ensuring the availability of medical records at any time.

Blockchain is a decentralized system that stores data and ensures its consistency. In [1], several components that are required to accomplish these functionalities are viewed as a system of six layers namely data layer, network layer, consensus layer, incentive layer, contract layer and application layer. Among all these layers, the contract layer is responsible for deploying smart contracts. In [2, 3], it is stated that smart contracts are self-verifying and self-executing codes which are modelled as a set of state response rules. Integrating smart contracts into blockchain can aid in performing a task in real time. Construction of blockchain powered smart vehicles that are equivalent to autonomous vehicles is pointed as a use case of smart contracts. In [6], a new consensus algorithm named Mixed Byzantine Fault Tolerance (MBFT) has been proposed to improve the processing speed of blockchain. It is largely inspired from PBFT algorithm. In PBFT, a node should receive permission from all the other nodes for entering the next phase of consensus. In [7], it is pointed out that PBFT might reduce the processing speed with the increase in the number of nodes. But in MBFT, the nodes are splitted into two consensus groups where the communication between the nodes are reduced. In [8], Elliptic Curve Digital Signature Algorithm (ECDSA) is integrated to achieve the consensus. Latency involved in running complex consensus algorithms can be reduced by ECDSA. In [9], a consensus has been generated and maintained in a decentralized manner. It helps in improving the resilience of the system and thus helps to recover from the system failure. Thus, the blockchain interacts with the decentralised record keepers to reach a consensus mechanism. Data Access are regulated using smart contracts. As blockchain is a decentralised system, data sharing is considered as atomic and intended entities are involved in the network. Smart contracts have the access of granting access to the doctors. It helps in improving privacy of the network.

In [5], Hawk, a smart contract system is proposed for securing the transactional privacy. Blockchain is used as a conceptual party which enhances the availability. It applies zero knowledge of correctness and proposes a universal composability framework model for the blockchain model of cryptography. In [10], a novel consensus named Proof of Block and Trade (PoBT) has been proposed for the integration of IoT with blockchain. As the IoT systems are lightweight, PoBT was designed to work. A ledger distribution mechanism has also been used to minimise the memory requirements of IoT nodes. In [11], various applications of blockchain in both non-financial and financial sectors have been discussed. Information has been provided on challenges and business opportunities in blockchain beyond cryptocurrency based technology. In [12], the positive implications of blockchain for modern organizations have been mentioned, specifically in the areas of financial services and physical asset ownership. In [13], Distributed ledger technology (DLT) has been used to minimize the cost of core verification procedure in financial institutions and to enhance the customer experience. Blockchain is applied in the supply chain which resolves the trust issues and trace records through norms. In [14], blockchain is used for converting the traditional centralized system to decentralized or mutli-centered system, to enable collaboration among different groups.

Blockchain is used in the field of medicine to maintain confidentiality of medical records [15–17]. Medical stakeholders participate in the network as blockchain miners. Contracts contain metadata about the record ownership, permissions and data integrity. Transactions in the system constitute instructions to manage these properties. State-transition functions of contracts have policies that are used to enforce data alteration only by legitimate transactions. This methodology is more adaptable at this situation because as the size of the dataincreases, the security needed rises too. The only drawback in this smart contract based method is that the patients doesnot have control over their data.

A distributed attribute based signature scheme for medical systems based on blockchain technique is proposed for securing EHR. This works well for small sets of data. But when the size of the data increases, scalability becomes a question. Blockchain technologies are expected to make a significant impact on a variety of industries. However, one issue holding them back is their limited transaction throughput, especially compared to established solutions such as distributed database systems. Here, a modern permissioned blockchain system has been rearchitected using Hyperledger Fabric in order to increase transaction thorughput from 3000 to 20000 transactions per second. The only drawback using this methodology is that the computational power required to process these data increases.

In [18] blockchain-based system was proposed to secure the medical information of patients, and thus improving consensus mechanisms for enhancing security and privacy. This is one of the best methodologies to provide securityand acheive scalability of the data. This is because, in order to enter any network, every other node must have their word on it, which makes it more secure. Since we need every node to accept a node to the network, the miner's digital signature, computational power needs to be high enough which may lead to system crash.

Blockchain-based architecture was proposed for e-health domain that has an enhanced mechanism to secure privacy details. It has consolidated the unique features of blockchain such as immutability, anonymity of users along with the modified classic blockchain structure. These modified design helps in overcoming the challenges of IoT applications such as higher overhead, lower throughput and latency. Thus, processing and cluster of miners, storage and process of data are performed at the nearby cluster of patients.

In [19, 20], cyberattack issues that arise among the healthcare systems were pointed out. Especially, EHR breaches were increased during the recent years. Hence, as the future predicts on increased security attacks on EHR, more researches were started on deploying solutions to ensure the security over health records storage. Mostly, integrity and consistency are considered as the vital characteristics to secure in EHR.

Traditional EHR, each medical services are managed with their own health records. These are found very difficult while the platforms are changed for sharing the records. However, price and size of blockchain remains a constraint while the data are stored in it. Therefore, cloud computing is considered for storing the data which can be shared in different platforms without any constraints. Data that are generated in the healthcare industry is increased enormously and hence the Healthcare Information Exchange (HIE) has formulated the rules for storage of data and securing the data. Important attributes (confidentiality, authentication and integrity of medical data) for securing the medical data were addressed and a secure EHR system with the Attribute-Based Encryption (ABE) and Identity-Based Encryption (IBE) were used for encrypting medical data along with the Identity-Based Signature (IBS).

3. Proposed Work

As health records are sensitive to store and handle, those records has to be maintained in a secured storage. When EHR is to be maintained in the blockchain, proper framework has to be selected in order to maintain and store as well. Also, EHR storage and sharing has to be provisioned with other doctors for the benefit of patient. Thus, our problem focuses on designing a platform where storing as well as sharing of medical records must be made possible with all the doctors whom the patients or the connected participants with the patients chose. Therefore, secret key is provided to the patient on connecting to the network, and the doctors selected from the patient's side are provided grant access to view and update the patient records. The framework and implementation details are discussed as follows:

3.1 Types of Blockchain

As the blockchain application differs according to the network size and consensus, frameworks are of various types. Based on the access permission, it is thus classified as

- Public
- Private
- Consortium (Hybrid)

(i) Public Blockchain

As the name suggests, anyone can enter into the network and have access to the block data. It has public Distributed Ledger Technology (DLT) where anyone can interact using the internet connectivity and can join as an authorized miner to mine a block. However, users identity is protected as a pseudo-anonymous hash value. Once the user has joined in the network, they are allowed to check and mine a block which are to be added in the network. These kind of public blockchain is used in financial incentives and PoW consensus is used. Examples are Bitcoin, Ethereum, Litecoin.

(ii) Private Blockchain

Private blockchain works based on access control rules that are distributed across the network and hence it is defined as a restrictive or permissioned blockchain. It is operated within an organization where one node controls the rules for making other nodes to perform smart contracts or miners. Security, permissions authorization and access permissions are controlled by organization. Examples are Hyperledger Fabric and Ripple. No participant can become a part of the network without any invitation from the controller.

(iii) Consortium Blockchain

Consortium blockchain is defined as partly centralized and decentralized and it is not used by single organization and it can be expanded to several organizations. It is accessed by a group of previously registered nodes which cannot directly access network without registration. It is used in the enterprises. Examples are Hyperledger Fabric, Quorum and Corda.

As our proposed Ethereum based Electronic Health Record storage system needs a permissioned public blockchain, the implementation uses Ethereum framework. The transaction details on Ethereum Network are as follows:

From	To	Value	Data	Gas	Gas Limit	Gas Price
20-Byte Address	20-Byte Address	Fund Amount (Ether)	Message sent to recipient	Transaction fees	Maximum Amount	Sender option to pay

4. System Architecture

EHR storage consists of three sections namely Data Users Layer, Data Management Layer and Blockchain Cloud. EHR collected from the hospitals or medical practitioners are uploaded into the cloud with their credentials. The role of blockchain is integrated with the data management layer where the processing of data are performed. In the proposed work, Ethereum and Inter-Planetary File System (IPFS) are used for storing medical records.

Smart Contracts are new buzz word which accompany the blockchain concepts. Smart contracts are the codes which are executed during the transaction phases. These are defined as smart contracts, as they can avoid security attacks such as tampering and alterations. It uses solidity language which can perform any kind of functionality in programmer's perspective. After programming the operations, it is then compiled using EVM bytecode. These contracts are build using Proof of Authority (PoA) which needs authority for executing each transaction. An interaction between the Ethereum Node and IPFS are as shown in Fig. 3. User act as an intermediate between the Ethereum node and IPFS. Once data is sent to the IPFS, hash content is sent to the user and the user store the content hash in the Ethereum Node.

In our proposed Ethereum based Electronic Health Record storage system, the patient wishes to store the electronic health records must register and upload the medical files and hash of the contents stored is returned to the respective patients. Thus, the data is encrypted and pushed into the Ethereum network. When the records has to be shared with the patients, doctor must be granted permission from the patients and hence

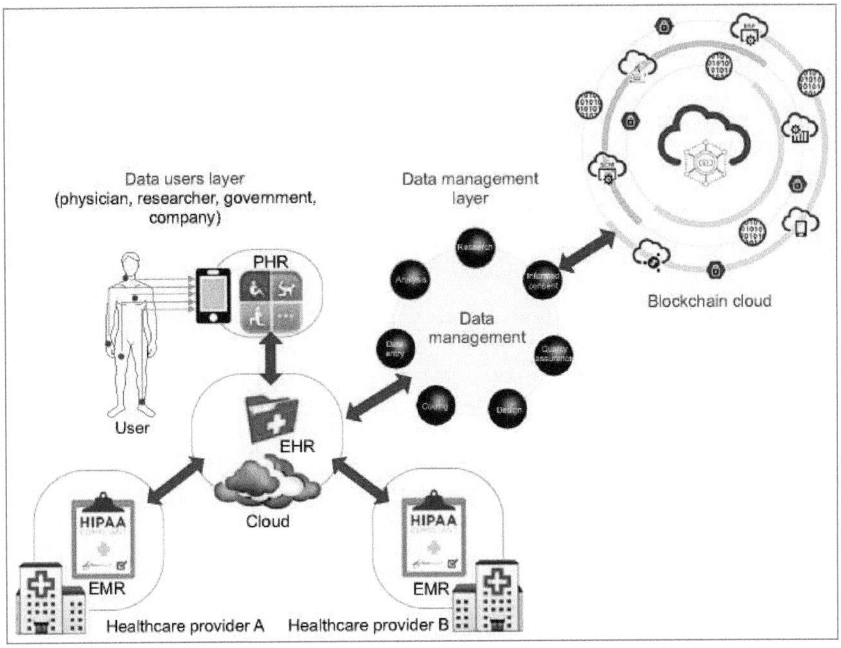

Fig. 3. Overview of storage of electronic health records.

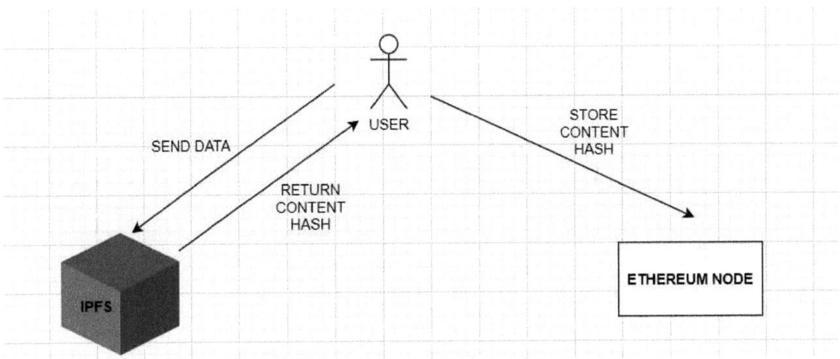

Fig. 4. Interaction between Node and IPFS.

the secret key will be used on providing access to the doctors. Once, the doctor is granted acces, he can view, update the records on requesting the file key. The flowchart of proposed mechanism is as follows:

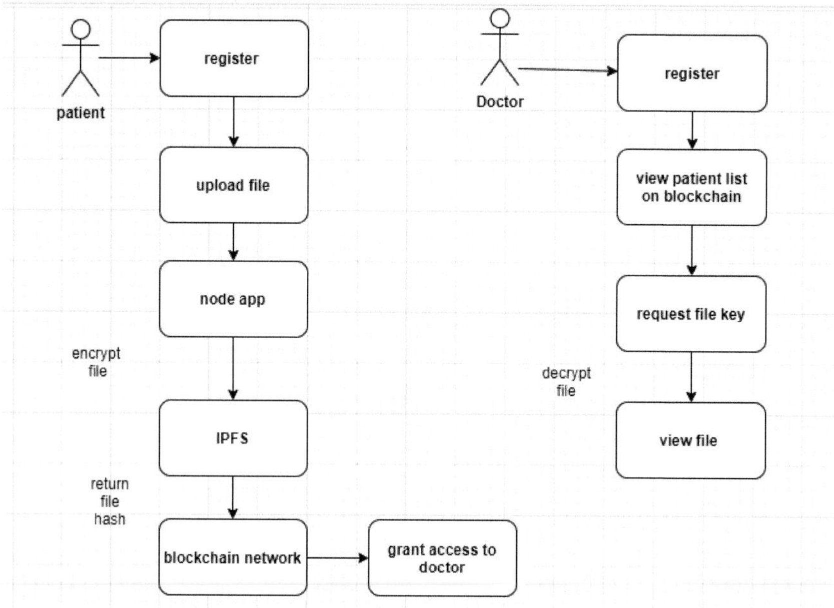

Fig. 5. Flowchart of proposed Ethereum based Electronic Health Record storage system.

Under the deployed contracts, the following features are available which are used for searching the respective patient records.

Add File	Patient's medical records are available and can also be added along with the IPFS hash content
View Records	Using the hash value, the records can be viewed. Only the authenticated persons can view their records
Grant Access	Doctors can be granted access to view the medical records
Signup	Patients who are not registered can use signup for storing the records
CheckProfile	Patients can be viewed with their address provided
getDoctorInfo	Doctors those who are connected can be viewed along with their details

The implementation uses Remix IDE and ethers get updated in the test Network called Metamask. These wallets are used for sending and receiving ethers on the network. It also have test networks in which we can have test ethers. Using those ethers we can execute the transaction on the network.

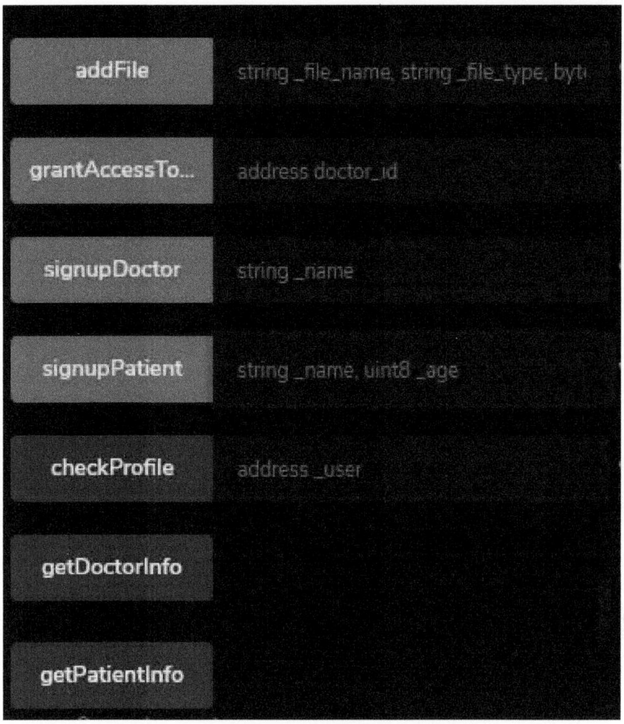

Fig. 6. Attributes of deployed contracts.

Algorithm:

checkPatient(msg.sender) check patient exist or not;
patient p = patients[msg.sender];
if(filehash.value=0)
  add (filename,file type,secret) to files;
  uint pos = p.files.push(*File Hash*);
function grantAccessToDoctor($doctor_i d$)
publiccheckPatient(*msg.sender*) *checkDoctor*($doctor_i d$)
  patient p = patients[msg.sender];
  doctor d = doctors[$doctor_i d$];
if ($p.doctor_l ist[doctor_i d] < 1$) *//doctor already have access*
$pos = p.doctor_l ist.push(doctor_i d)$;
$d.patient_l ist.push(msg.sender)$;
  function getFileInfo(fileHashId)private view checkFile(fileHashId)
  returns(filesInfo)
  return hashToFile[fileHashId];

```
  function getFileInfoDoctor(doc, pat,fileHashId) public viewonlyOwner
  checkPatient(pat)
  checkDoctor(doc) checkFileAccess("doctor", doc, fileHashId,pat)
filesInfo memory f = getFileInfo(fileHashId);
return (f.file_na)
```

Fig. 7. Transaction on the Ethereum Netwo.

5. Result

Thus, the healthcare system for storing the EHR was designed in Ethereum network and the smart contracts are defined in the solidity language. Transactions are executed using the attributes discussed and the doctors are provided access using the secret key.

Protecting the confidentiality, integrity and availability of a patient's data is subtle and vital among threats.

Figure 8 shows a sample risk matrix that helps in analyzing the threats. This risk matrix will be used to explore the various metrics of

RISK PROBABILITY			RISK SEVERITY		
	EXTREMELY HAZARDOUS A	HAZARDOUS B	MAJOR C	MINOR D	NEGLIGIBLE E
FREQUENT 5	5A	5B	5C	5D	5E
LESS FREQUENT 4	4A	4B	4C	4D	4E
REMOTE 3	3A	3B	3C	3D	3E
SLIGHTLY IMPROBABLE 2	2A	2B	2C	2D	2E
EXTREMELY IMPROBABLE 1	1A	1B	1C	1D	1E

Fig. 8. Risk matrix for analysis of the system.

the systemThe designed and implemented system works in a multi-flowed way. The distributed architecture eliminates the risk of a single point of breach. This ensures that the integrity of the data is protected in case of a breach at one of the distributed storage. If a normal centralized system suffering from a single point of breach and has the risk rating of 3B assuming the data isn't backed up, the designed system reduces the risk rating to 3D since the integrity can always be verified by checking the data in the affected point of breach with other distributed copies.

Consensus mechanisms are fundamental feature of blockchain helps to satisfy the integrity of the shared storages.Without the consensus of the participants, any ledger or a new block cannot be manipulated, Hence the risk of a threat exploiting a vulnerability in a single storage is reduced. In the normal healthcare system, if the risk of a particular storage being corrupted by a threat that has the risk rating of 4C, the designed system reduces the risk to 2C.

6. Conclusion

This chapter provides an in-depth knowledge about blockchain and its evolution. Importance of storing EHR is then explained along with the taxonomy features in Section 2. Related works on blockchain frameworks were summarised. The proposed work on, securing of EHR is analysed and implemented in Ethereum framework, which consists of patients and doctors connected in the network. Framework details and the implementation are discussed in Sections 3 and 4. Results of the proposed work are explained in Section 5 along with the risk analysis. Hence, the proposed work provides security on storing EHR with increased efficiency. As a future work, the proposed work is to be extended with the different medical organizations for which Hyperledger framework will be beneficial. It can also be extended to IoMT, cloud and edge computing.

References

[1] Nakamoto, S. 2008. Bitcoin: A peer-to-peer electronic cash system. Decentralized Bus. Rev., 2008, Art. no. 21260.

[2] Vazirani, A.A., O'Donoghue, O., Brindley, D. and Meinert, E. 2019. Implementing blockchains for efficient health care: Systematic review. J. Med. Internet Res., 21(2): 2019, Art. no. e12439.

[3] Hölbl, M., Kompara, M., Kami²ali¢, A. and Zlatolas, L.N. 2018. A systematic review of the use of blockchain in healthcare. Symmetry, 10(10): 470.

[4] Agbo, C.C., Mahmoud, Q.H. and Eklund, J.M. 2019. Blockchain technology in healthcare: A systematic review. Healthcare, 7(2): 56.

[5] Chukwu, E. and Garg, L. 2020. A systematic review of blockchain in healthcare: Frameworks, prototypes, and implementations. IEEE Access, 8: 21196–21214.

[6] Sun, Y., Zhang, R., Wang, X., Gao, K. and Liu, L. 2018. A decentralizing attribute-based signature for healthcare blockchain 2018 27th International conference on computer communication and networks(ICCCN) (2018), pp. 1–9.

[7] Gorenflo, C., Lee, S., Golab, L. and Keshav, S. 2019. FastFabric: Scaling Hyperledger Fabric to 20,000 Transactions per Second 2019 IEEE International Conference on Blockchain and Cryptocurrency (ICBC)(2019), pp. 455–463.

[8] Fan, K., Wang, S., Ren, Y., Li, H. and Yang, Y. 2018. Medblock: Efficient and secure medical data sharing via blockchain. J. Med. Syst., 42(8): 136.

[9] Thakkar, P., Nathan, S. and Viswanathan, B. 2018. Performance benchmarking and optimizing hyperledger fabric blockchain platform 2018 IEEE 26th International Symposium on Modeling, Analysis, and Simulation of Computer and Telecommunication Systems (MASCOTS) (2018), pp. 264–276.

[10] Chen, L., Lee, W.K., Chang, C.-H., Raymond Choo, K.-K. and Zhang, H. 2019. Blockchain based searchable encryption for electronic health record sharing. Fut. Gener. Comput. Syst., 95: 420–429.

[11] Chen, L., Lee, W.K., Chang, C.-H., Raymond Choo, K.-K. and Zhang, H. 2019. Blockchain based searchable encryption for electronic health record sharing. Fut. Gener. Comput. Syst., 95: 420–429 2019 IEEECanadian Conference of Electrical and Computer Engineering (CCECE).

[12] Jiang, S., Cao, J., Wu, H., Yang, Y., Ma, M. and He, J. 2018. Blochie: A blockchain-based platform for healthcare information exchange. 2018 IEEE International Conference on Smart Computing (SMARTCOMP),pp. 49–56.

[13] Xia, Q., Sifah, E.B., Asamoah, K.O., Gao, J., Du, X. and Guizani, M. 2017. Medshare: Trust-less medical data sharing among cloud service providers via blockchain. IEEE Access, 5: 14757–14767.

[14] Wang, H. and Song, Y. 2018. Secure cloud-based EHR system using attribute-based cryptosystem and blockchain. J. Med. Syst., 42(8): 152.

[15] Guo, R., Shi, H., Zhao, Q. and Zheng, D. 2018. Secure attribute-based signature scheme with multiple authorities for blockchain in electronic health records systems. IEEE Access, 776(99): 1–12.

[16] Zhang, J., Xue, N. and Huang, X. 2016. A secure system for pervasive socialnetwork-based healthcare. IEEE Access, 4: 9239–9250.

[17] Zhang, X. and Poslad, S. 2018. Blockchain support for flexible queries with granular access control to electronic medical records (EMR). 2018 IEEE International Conference on Communications (ICC), pp.1–6.
[18] Uddin, M.A., Stranieri, A., Gondal, I. and Balasubramanian, V. 2018. Continuous patient monitoring with a patient centric agent: A block architecture IEEE Access, 6: 32700–32726.
[19] Cohen, I. Glenn and Michelle M. Mello. 2018. HIPAA and protecting health information in the 21st century. Jama, 320.3: 231–232.
[20] Choi, Young B. et al. 2006. Challenges associated with privacy in healthcare industry: Implementation of HIPAA and the security rules. Journal of Medical Systems, 30.1: 57–64.

Chapter 9

Secure IoT Smart Health Monitoring System (SHMS) for Patients using Blockchain Technology

G Indra Navaroj,[1,*] *M Kavitha Margret*,[2] *I Solomon*[1] and *B Valarmathi*[3]

Internet of Things (IoT) technology is very hot area in recent days. The technology is now apply many of the applications like smart home, smart parking system, smart metering, smart health care and smart agriculture. Smart Health Monitoring System (SHMS) deals with the patient health condition and monitoring remotely while they needed the immediate treatment. That health monitoring system is continuously monitoring the patient health; communicate to the doctor, nurse and patients. That is very useful for the patient and also the doctors to reduce the tension, death rate and the valuable time also. Health Monitoring System (HMS) contain three functional layers such as Information Collection layer, System Deployment layer and Service Dissemination layer,

[1] Jayaraj Annapackiam CSI College of Engineering, Nazareth, Tamilnadu.
[2] Sri Krishna College of Technology, Coimbatore, Tamilnadu.
[3] Vellore institute of Technology, Vellore.
Emails: kavithamargret.m@skct.edu.in; king23rdsolomon@gmail.com; valarmathi.b@vit.ac.in
* Corresponding author: indrajesus@gmail.com

that layers are used to perform the operation of the patient health. Many number of IoT devices are connected to the patient body and measure the patient body conditions (health). The system is provides the alert ration to the patient' relatives and the doctors. So easily save the patient life and avoid the unwanted death. In this chapter, proposed Blockchain Health Monitoring System (BHMS) used to monitor the patient heartbeat, temperature, blood pressure and sugar level. However, security is a major problem for sharing patient health history, insurance claim and pharmacy supply chain. In this chapter, concentrate the security of health monitoring system. BHMS provide the security of patient's historical information.

1. Introduction

In very first time kevin asthon was coined the terms Internet of Things. In IoT all the things may be physical and virtual are communicate and accessible trough internet. Different tings arrounds us are to be internet worked. Internet are basically a global network or an internet work of different computers and computing devices. Computing and computer devices were connected. Interconnect different things that means physical object around us such as lighting system in a room, air conditioner, fan, fridge, anything including things such as microwave oven, tooth brush, refrigerator and so on. Not only in home things in business are things such as machine, different equipment internet working. Different things lighting system, chair, table, watch are fitted with embedded system, information technology embedded electronic are provide computing platform [1].

IoT is one of the building blocks that are considered to be of use for developing smart homes and smart cities. IoT is one of the enabling technologies to make the city smart, to make the home smart. Different things are connect in the internet are increased rapidly. Large numbers of nodes are connected internet worked. The things that are connected to the internet are going to be projected to cross the 20 billion in the future. Internet work of things is not a single technology. Physical devices can be of different types of physical devices having different configurations, different specifications and so on [2]. Each of these supported through different other technology such as cloud technology, big data machine and learning networking computer vision. These different technologies are from electrical sciences and some from mechanical sciences that are required in order to build IoT. IoT support new era of ubiquity, so anywhere, anytime, anything are communicated and connected to the internet.

Now all human being are also connected to the internet, so all these different things and all these different devices are send lot of data. Some areas are identified as IoT enabler RFID, sensor, nanotechnology, smart network. RFID technology has RFID reader, RFID tags, which build Internet of Things. Sensor is the important enabling technology or enabling device that builds Internet of Things [17]. Very small size nano devices are going to be used; they are going to be allowed and consumed and then know once that is done in the form of capsules, they are going to be internetwork. Networking of physical object ate communicate, sense, interact to the external environment [3]. Each and every things have the capability to sense the surrounding, perform operation and communicate to each other.

The nature of the IoT things or object are able to collect the data from the surrounding and transmit the data to the private and public to compute the important data then ensure the security of the data. And also maintain the integrity and confidentiality of the transmitted data and provide the authentication of the sensitive data [1]. The important components of IoT are sensing, heterogeneous access, information access, services and access. Another important components are security and privacy [2]. Internet of Things describe daily used object from home or office, anywhere will be interact and interconnected with their surrounding in order to collect and control the element and perform automatic task. Then concentrate on other things like security, privacy, seamless authentication, attacks, self-maintenance and easy deployment [4]. Important building block of IoT technology is sensor and actuator. Sensors are sense the physical property around us occurring and actuator are based on the sense information and perform the operation on the physical environment.

They sense different parameters depending on the sensor being used. For example, temperature, pressure, humidity conditions, lighting conditions and so on. Then, what will happen is these sensed information are going to be sent over a connected system [5]. That means, over a network that information will be passed, it can also involve cloud. That information is going to be transmitted based on what has been sensed and based on the requirements, some physical action is going to be taken by an actuator [15]. Many of author used Rasperry pi board to monitor the patient's heart beat, body temperature, respiration and body movement. Smart health monitoring system is an important application in smart cities that system allows the doctor to monitor patient health condition and give the treatment in remotely [16].

1.1 Health Monitoring System

Medical services are very important for the each and every nation. Our country spends lot of money for the welfare of people and children.

Due to increase the population and the disease we need the additional hospital and doctors and nurse. It is increase the cost and lack of doctors and nurse. So the modern medical system is support the Tele-health or e-health. Tele-health or connected health use information and communication technology to remotely monitor patient health and doctor provide medicine care for patient. Many of research is being done in the field of Tele-health. The use of ICT to provide the care based on the information given by the patients. The information is based on the surrounding information [18].

Ambient Assisted Living (AAL) platform is popular for health care service that provides automated health service. The wearable cloud is introduced which collect the information and provide interface with patient and doctor. SHMS offers peace of mind; reduce treatment cost and travelling time. This chapter proposed the secure Blockchain health monitoring system that supports the patient medical care. That is provides the security and privacy of the patient medical information or sensitive data. This proposed work using the Blockchain technology for the architectural design of the secure health monitoring system.

2. Related Works

Kumar et al. [6] monitor patient temperature, heart beats, patient movements, respiration based on Rasparri Pi linux operating system. Sensor senses the signal and sends the signal to the rasperri pi through Amplifier circuit and Signal Conditioning Unit (SCU). If the signal level is low, amplifier is used to increase the signal strength and send to the Rasperri pi. Rasperri pi act as a processor. Then the patient temperature, heart beats, respiration and patient movement measured respective sensor and monitor computer screen. Author uses the thermistor to measure the patient temperature. Thermistor is a temperature sensitive resistor. IR transmitter and receiver are used to monitor the patient heart rate. If the patient heart rate is above the 100 bpm, send the alert to the doctor or nurse. There are two thermistor are used to measure the patient respiration, one is used for respiration another one is used for room temperature.

Gao et al. [7] developed IoT health monitoring system for measure the pulse rate and saturation peripheral oxygen then collect and analyze data using the IoT. Collected data send to the central cloud server. Author develops the IoT health monitoring system using android application development system and Amazon cloud server setup. Android application is used for collect the data, it act as a bridge between the central cloud server and measuring device. Server is used for Access, manipulation, data storage, analysis. IoT health care system consist of three component which are a Nonin 3230 Bluetooth® Smart Pulse Oximeter is used for measuring pulse rate and peripheral oxygen saturation and, a Google Nexus 9 Tablet is used for data transferring , and an Amazon Web Services (AWS) central cloud server is used for manipulation and data storage. This IoT system, the oximeter and the tablet are used to a patient assigned, then doctors and nurses can access the patient data through a LAMP (Linux+Apache+MySQL+PHP) query system on an AWS server. This IoT health monitoring system can be used in the real time patient care.

Nidhya et al. [8] develop the health care system for the convenient of doctor and patient without their presence. So we need the security of the patient information. It meets the vulnerability of privacy data. Author propose a novel end to end energy aware routing architecture for securely monitor the patient health condition BSN health care server using the anonymous lightweight authentication protocols to provide the security in modern health care, and energy aware routing achieve efficiency and end-to-end trust in IoT healthcare applications. In Body Sensor Network, consist of collection of body sensor in and around the human body that is used to sense and monitor the human body condition. Then collected data are sends to the local processing unit, it is also called central processing unit it act as a bridge between the BSN node and Local Processing Unit (LPU). Author develops the security and energy efficient of Patient private data.

Adam [9] proposed the Remote Monitoring and Management of platform of Healthcare Information (RMMP-HI) can used to monitor and manage the patient health and diseases. Main purpose of RMMP-HI is provides prevention and early detection of diseases. Due to restriction of time and location RMMP-HI can collect the medical information timely with the help of body sensor. Many numbers of sensors is used to collect the surrounding information and continuously monitor the patient condition. Sensor is used to collect the useful and valuable information, store the information, compute the operation and perform the comparative analysis. When the body sensor found abnormal information give the alert to the patient take the treatment to early prevent the disease. It contain national health monitoring record, it is provide gather medical information and decision making, analyzed related records and processing healthcare information.

Kajomkasirat et al. [10] health monitoring system collects the patient information through the wearable devices using API technology. Android studios are used for mobile application. The system was developed to give health information to the patient through mobile application. Author used MYSQL, JAVA, HTML5, PHP script for design the database. Health monitoring system is maintains the patient profile and give the alert message to the user. Wearable sensor is used to monitor the human activity.

MYSQL is used to collect the health information data is analyzed and give the health recommendation to the human. Author used many of the technology like MYSQL technology, PHP technology, JAVA technology, Android technology, Bootstrap technology, Cloud technology, Data Mining technology. Rapid Miner is used for analyze the collected data and give the alert to the human.

Simić et al. [11] used to monitor the cancer patient health, business analytic, cloud service, implementation of WSN and IoT technology for smart health care delivery. That is used to improve the quality of life and improve the safety and security of cancer patient. Banak et al. [12] propose an automatic system to collect and monitor patient's heart beat, body temperature, blood pressure, body movement. The system is predicts any symptoms about the diseases to recommend the needed treatment about the patient. Author developed smart health monitoring system using Rasperry pi. The system is continuously monitoring the patient's body temperature, blood pressure and heart beat. Collected information stored on cloud storage and accesses the information with the help of mobile application by only authorized users.

3. IoT Health Monitoring System Overview

In human health is a primary element for each and every person life. Health is used for human to feel good and act effectively and perceive. Health describe body health, mental health, spiritual health, social health [13]. Health care system, device that monitor the patient state and measure the state of and give the information to the doctor through internet can solve the major problem. It is used to solve the patient health problem, protect the patient health and save the doctor time. In smart Health Care System use wearable device, all the data is directly recorded so doctor still need to examine the patient. Doctor checks the patient, he want to write all measured data. Sometime doctor forgot to write measured data and write something wrong [14].

If doctor check the patient body condition then doctor found some problem in patient health then doctor verify all patient health history. These wearable devices are used to continuously monitor the condition and the device is connected to the internet. If the patient is very serious doctor save the patient and save valuable time and refer the measured data from these devices. Connected device continuously monitor the patient health 24/7 hours and provide valuable data to the doctor. In Internet of Things (IoT) technology, immutability is used to patient valuable data is not change by the third party and preserves the complete history of patient. IoT devices collect huge amount of data and automatically saved in patient Electronic Health Record (HER).

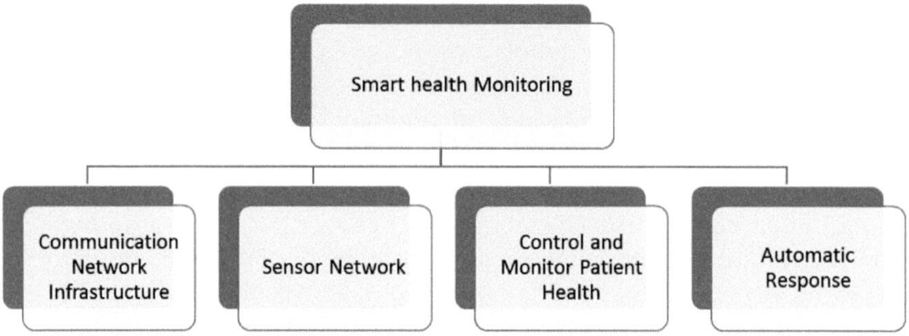

Fig. 1. Structure of Health Monitoring System.

The structure of IoT health monitoring system consist of four major components that are communication network infrastructure, sensor network, control and monitor patient health and automatic response to patient.

Communication network infrastructure consist of the communication technology like IEEE 802.15.4 standard, Zigbee, 6 LoWPAN, Wireless HART, Z-Wave, ISA 100, Bluetooth, NFC and RFID. That communication technology is providing the communication of information. All the things are communicated via internet.

Sensor network is based on IoT technology and Wireless Sensor Network (WSN) technology. It consists of the component of sensor, actuator and wearable device, that device is used to monitor the patient health. If needed any help from the doctor or nurse give the alert to the doctor and nurse. Control and monitor the patient's health is continuously monitor the patient's heart beat, respiration, body movement and also perform task. Automatic response describe take automatic response to the patient, doctor based on the diseases and treatment.

Another major problem of health monitoring system is how to choose the doctors. More number of doctors is available in health care system. If the patients are new born baby, so need the pediatric doctor and the patient is cancer affected then need the cancer care center and patient is heart patient then need the heardiatric. So need to overcome the security problem remotely. Then the propose system want to support to pharmacy supply chain also. The needed pharmacy is sent to the corresponding patient home or hospitals.

4. Backgrounds of BHMS

Blockchain and attribute-based cryptosystems can support secure Electronic Health Records (EHRs). Identity-Based Encryption (IBE) and Attribute-Based Encryption (ABE) encrypt medical data. Identity-Based

Signature (IBS) implements a digital signature (Wang and Song 2018). A smart contract is used to handle Protected Health Information (PHI). A Blockchain-based smart contract is implemented to manage medical data and secure analysis (Griggs et al. 2018). Several solutions have been proposed to manage patient health data using a distributed ledger. The system includes the provider, hospital, medical insurer and patient. It provides the solution of clinical issues; all patients are trusted because the private Blockchain allows for security and privacy of sensitive patient information (Chakraborty et al. 2020). The health chain in Blockchain technology is to enhance medical devices. A patient can add or remove a doctor from the health chain (Xu et al. 2019). Automated doctor diagnoses and patient health data cannot be modified or deleted. Biosensors measure and collect patient information. In addition, the system is responsible for storing the patient's health status to Blockchain smart contracts (Dey et al. 2017).

5. Elements of Health Monitoring System

Health Monitoring element contain the various type of sensors. These sensors are used to measure or monitor the patient health.

1. Pulse Rate Sensor
2. Body Temperature Sensor
3. Body Position Sensor
4. ECG Measuring Sensor

Pulse Rate Sensor

Pulse sensor Amped is implemented to measure the patient heart pulse rate information. Patient pulse rate information can be helpful for planning a practice schedule and movement of patient body. It is put in your shirt to flicker. The challenge is that heart rate can be hard to measure. Pulse sensor Amped can solve these challenges.

Body Temperature Sensor

The DS18B20 is the best temperature sensor for measuring the patient body temperature. It contains three wires like GND, VCC, and DATA. Temperature sensor put at different place of human body like under arms, under the tongue and on forehand.

Body Position Sensor

The body position sensor are the accelerometer ADXL335 is implemented to identify and differentiate the position of patients. These sensor is identify weather the patient sleep or not, walking, sitting, jumping and at front bend. That exercise is used to reorganization of monitor the patient health.

ECG Measuring Sensor

The AD8232 Heart rate monitors sensors are implemented by analyze the Electromyogram and Electrocardiogram. ECG sensor can be put at different spot of the human body like on hands, on chest, etc.

6. IoT Health Monitoring System

IoT technology is connected all the things and make the world smarter. IoT is contain number of connected, sensor and connected technology, then measure the surrounding environment, analysis and decision making. Multiple systems interact that make global IoT. The proposed SHMS is provides the solution for IoT application Health Monitoring System. That provides the new vision of health care system and integrated with various IoT applications. IoT is still backbone of various IoT applications like health monitoring system.

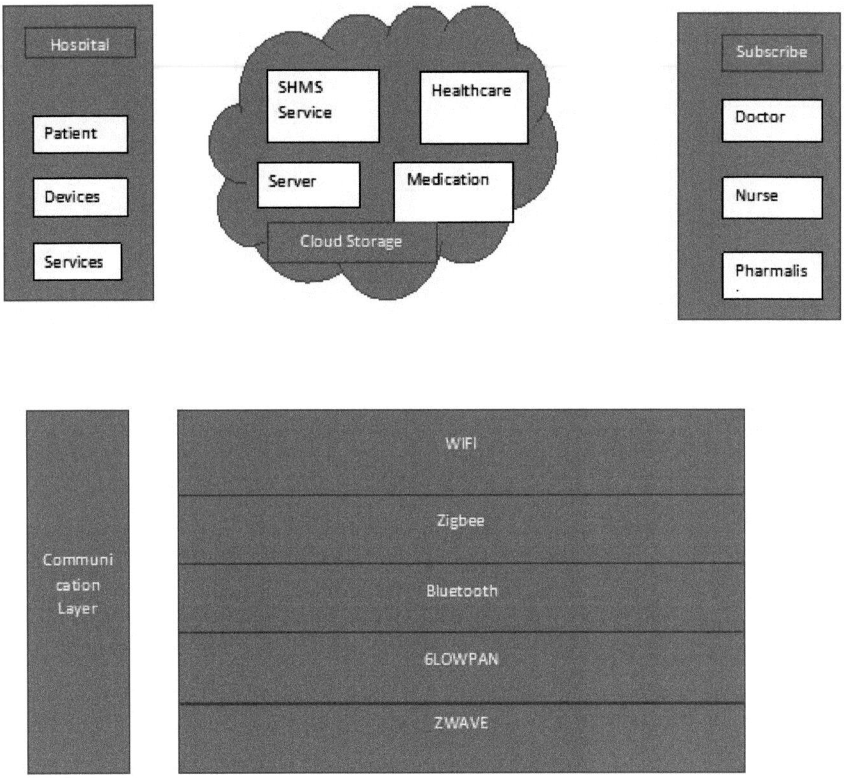

Fig. 2. IoT Health monitoring system architecture.

Vital sign service can be implemented for store the patient medical information. Each device performs the various functions that measure the patient data and store the data. For example, temperature service which receive the body temperature service from temperature sensor. It receive the data from patient and store the data then store data is sent to the analysis service. Analysis service is analyzing the received data for further treatment.

Patient service and subscriber service are responsible for patient medical information. The subscriber contains the different specialist doctor, nurse, loved one and care one. Doctor accesses the patient medical data and also gives the treatment. Based on the stored information the doctor diagnosis the symptom of the disease and give the treatment for save the life of the patient and also give the good quality of service. Different type of communication technology is used to communicate the patient to cloud storage and cloud to the doctor.

7. Blockchain Health Monitoring System

In centralized cloud storage everyone edit on their local copy of the patient sensitive data, treatment and medicine. So the Patient's historical or sensitive information are attack, modified the third party. However the patients met lot of critical issues. So, now concentrate the Blockchain technology for apply the IoT health monitoring system.

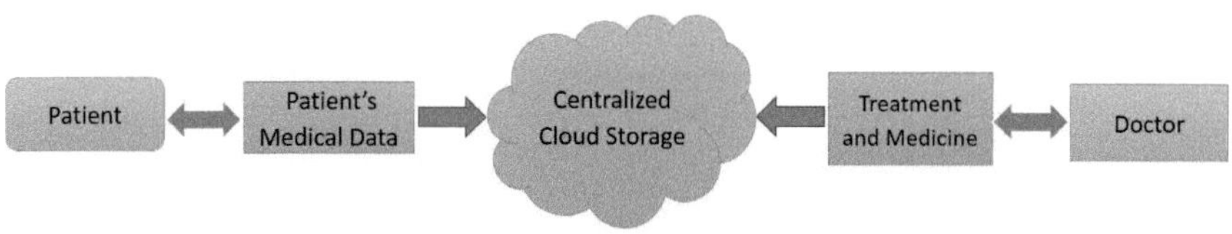

Fig. 3. Centralized cloud storage of patient's data.

Blockchain is a decentralized database with support the strong consistency of patient and doctor. Blockchain is provides the strong security of health monitoring system.

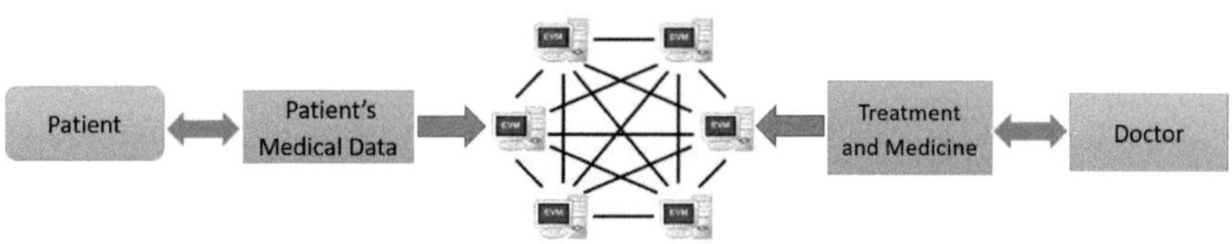

Fig. 4. Blockchain storage of patient's data.

In Blockchain network every patient maintains the local copy of the global information. So the system ensures the consistency of the local copy of the patient information. The local copy at every patient is identical. The local copies of the sensitive information's are updated based on the global information. In Blockchain network public ledger is a database, which maintains the historical information of every patient. The historical information might have been utilized for future treatment. In Health Monitoring System, the old treatment used to update the new treatment. Public ledger is used to store 500 patient's information.

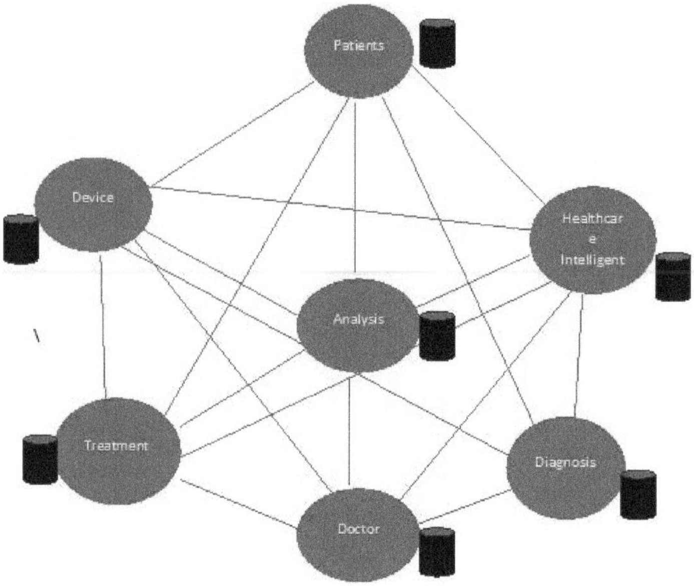

Fig. 5. Blockchain network.

BHMS support the following aspects:

Protocol for Commitment: System ensures that every patient's medical information from the patients are committed and included in the Blockchain with a finite time.

Consensus: System ensures the patient's sensitive data are consistent and updated.

Security: The patient's data need to be tamper proof. So the patient's may not act maliciously or cannot be compromised.

Privacy and Authenticity: The patient's medical information or sensitive information belongs to various clients. Blockchain ensure the privacy and authenticity.

Cryptograpically secured chain of Block

Hash function is used to timestamp a digital patient's medical data. Whenever a patients access medical information, construct a block consisting of the sequence number of access, patient's ID, timestamp, a hash value from the previous request and the entire data is hashed to connect to the previous block.

Merkle Tree

Merkle tree also known as hash tree. In 1992 Bayer, Harber and stornetta used Merkle tree for time stamping and verifying a digital document. The efficiency of improvement by combining time stamping of several patients' data in to one block, that block is called patient block. Merkle tree is used in the patient document, then the shared patient's data are unaltered, no one can lie about the patient document. That tree is used to patient blocks received in undamaged and unaltered, then other patient do not lie about the patient block.

Blockchain 2.0 is trying to use in industry, manufacturing, supply chain, finance, governance, IoT. Blockchain is a powerful technology. It is a decentralized platform can be utilized to avoid intermediate.

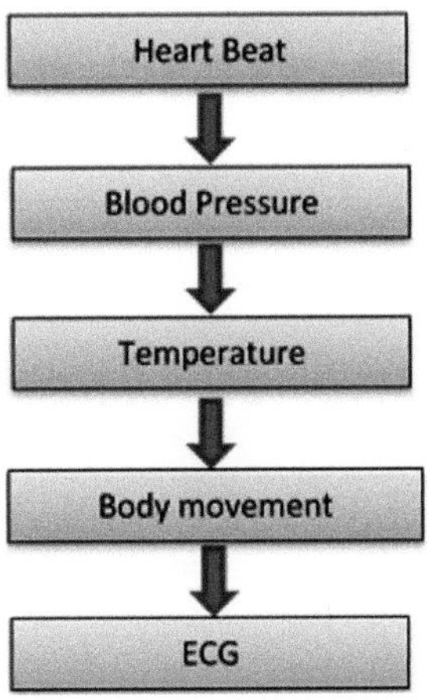

Fig. 6. Merkle Tree for BHMS.

Smart contract

Smart contract is a digital contract that is faster, cheaper and more secure. It provide legal contract, computerized protocol used for digitally facilitating, verifying, performance of legal contract by avoiding intermediates. It support to directly validating the contract over the decentralized platform. Smart contract support the crowd funding platform of health monitoring system. Patients access a medical information, construct a block consisting

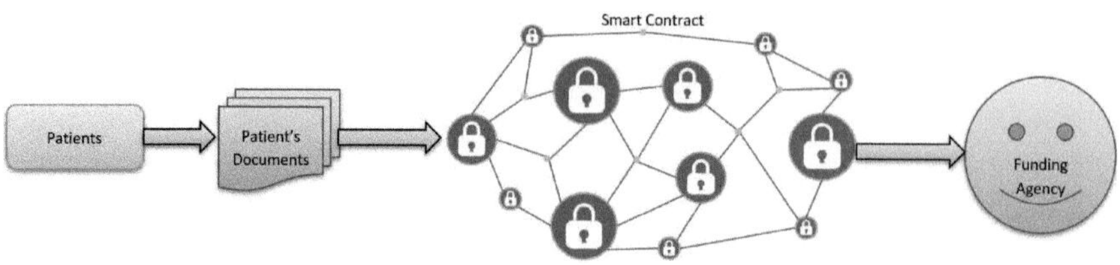

Fig. 7. Smart Contract for BHMS.

of the sequence number of access, patient's ID, timestamp, a hash value from the previous request and the entire data is hashed to connect to the previous block.

The patients have a 1disease, he want to take treatment in multi-specialty hospital, but do not have sufficient money to take treatment. In Blockchain smart contract support to submit the proposal in a crowd funding agency. Multiple agency commit to support medical treatment with small fund. In Blockchain ensure that patient get the complete money of the medical treatment is successfully completed.

8. Secure BHMS

In IoT based health care system use of sensor for monitoring patient health. Many of author did several research for health monitoring system.

1. Body Sensor
2. Central Node
3. Short-range Communication
4. Long-Rang Communication
5. Blockchain Architecture.

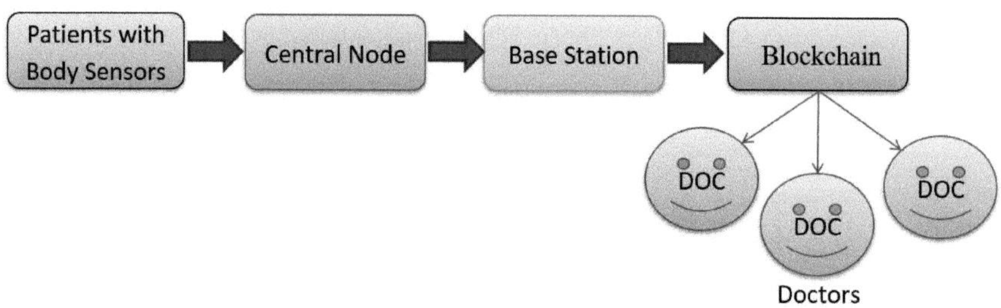

Fig. 8. Blockchain architecture.

Body Sensor

Body sensor contain many of the sensor such as the blood pressure sensor, body movement sensor, temperature sensor, body movement sensor, temperature sensor, blood sugar sensor, ECG sensor, Heart Beat sensor. That sensor is used to measure patient physiological condition. Body sensors are those that measure pulse, body temperature, respiratory rate and vital sign. These data are essential for determine patient critical health. Special purpose sensor like fall detection, blood –glucose could also be implemented for a specific condition. Another important sensor is used to measure blood oxygen sensor and blood pressure.

Central Node

Central node is act as a database. It is used to store the data from the body sensor. It is receive the data from the body sensor. Central node process this information, it analysis the data implement some decision making Blockchain network.

Short-range Communication

A short-range communication is required for sensors to communication with the central node. When choosing the short range communication, we consider the standard, security, latency and effect on the human body. It might provide strong security mechanism for the patient sensitive data. If you choose the method not provide any negative effects on the human body such effect may cause additional health concern for patients. The sensitive data cannot be attack by any attacker. Low latency is essential for short range communication, if needed by ambulance any critical condition of patient health give the alert to the alarm or call the ambulance. This is very important for health care system. Time delay could be the different between life and death.

Long-Rang Communication

Central node's data should be forwarded to a database where the doctors are sent securely access the data. Long Range communication are implemented message is transfer from one to another. The message sent by central node, the same message is received by the doctor that is ensured. It supports the high availability, which implemented message deliver at anytime, anywhere. In time critical application this is importance.

Blockchain Architecture

Medical information received from patient through body sensors must be stored secured for continuous use and future usage. Doctor is benefit from identifying a patient medical history. Blockchain are implements the security of patient information by the method of tamper proof. Blockchain provide different cryptography algorithm and hash function to securely store data.

9. Security in Blockchain Health Monitoring System

Devices Security describes the production of embedded device like body sensor, temperature sensor and body wearable device. Also provide the security of fog devices are gateway, processing unit, and hub.

Communication Security describe the security among the communication device like h1ealthcare device to fog device, fog device to fog device, fog device to cloud. Lightweight cryptography, cryptography algorithms are used for provide security in communication devices.

Data Production provide the data production among device data, communication data, cloud data.

Security Management is global security handling at cloud storage. SDN based security management and monitoring.

10. Potential Use Case for the Proposed Model

First use case is patient consent and health data exchange. If the patients are continuously gone to the particular hospital they have the medical history for the hospital. In case the patient went to the hospital to meet any doctor they access the patient medical history immediately. That person got in emergency situation or met the accident then rush in to the nearby hospital. There is no information about the person such that the person may be diabetic or not. Some of that kind of information is very critical and the kind of medicines you can prescribe to that persons. How IoT can make an impact is a health data exchange. In the hospital can talk to each other to collect the patient information. But the patient information is maintains very securely, that private information about the patient rights.

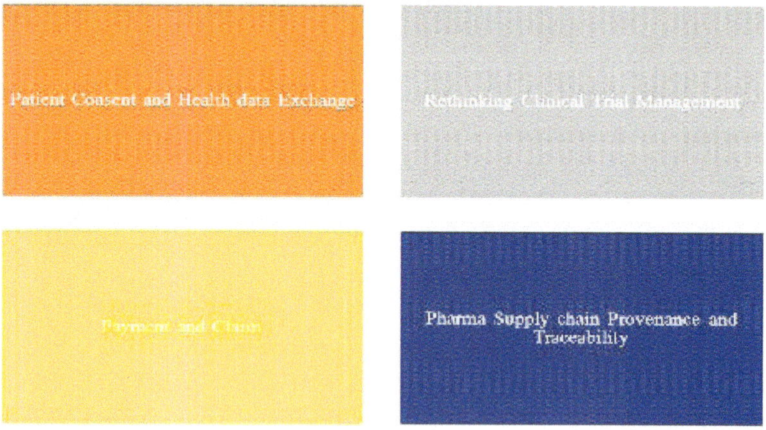

Fig. 9. Use cases.

Then it requires patient consent or the consent of someone high authorize and only after they provide that consent can the data exchange. The process of consent and data exchange can be something that can be captured on an IoT and because this is all private information is highly secure and needs to be audited, IoT provides that capability. So the patients feel secure that the patient information regarded and disclose their information.

Second use case is payment and claim. Now a day healthcare is very closely related to health insurance and involved lot of information. When the patient went to the healthcare, the health care gets the charge from the patient. If the patients have health insurance from the insurance company, they transfer the amount to the particular health care. Multiple parties are involved in this ecosystem so fraud is big part of this healthcare. So IoT support automatic claim processing and payment.

Third use case is pharmacy supply chain. Supply chain has lot of complaint aspect, traceability aspects. In IoT technology follow strict rule for a pharmaceutical codes have to be transported across some of them maybe refrigerated and so on. There is compliance or business rules are given how to process the supply chain and we want to able to trace it. A big problem of supply chain is a very relevant to India as well is counterfeit.

Fourth use cases are clinical trial management. Day today information is getting exchanged between multiple entities on clinical trials. About pharmaceutical industry maybe hundred different maybe molecules or whatever they use in a drugs maybe only one of those hundred trials might actually be even they might even take it.

11. Experimental Results

In this section, discuss some of experimental result of patient body health. BHMS is used to measure the five patient heartbeat, temperature and sugar level. Then produce the comparison chart for the following patient. So doctor and nurse easily analyzed the patent condition and give the suggestion, alert, treatment to the particular patient. Smart health monitoring system is used to monitor the patient health condition. BHMS is not only monitoring the patient health condition also provide the medicine and treatment to the particular symptoms or diseases. Many numbers of doctors are available in the BHMS to give treatment to them. It takes the care of different kind of diseases and cures them.

The patient's report is very useful for both patient and doctor for give the treatment immediately. If the doctor fined the patient health condition is very critical, continuously monitor the particular patient health. And also produced report chart for the particular patients. For example the patient Alice is very critical so the body's sensor and wearable device are

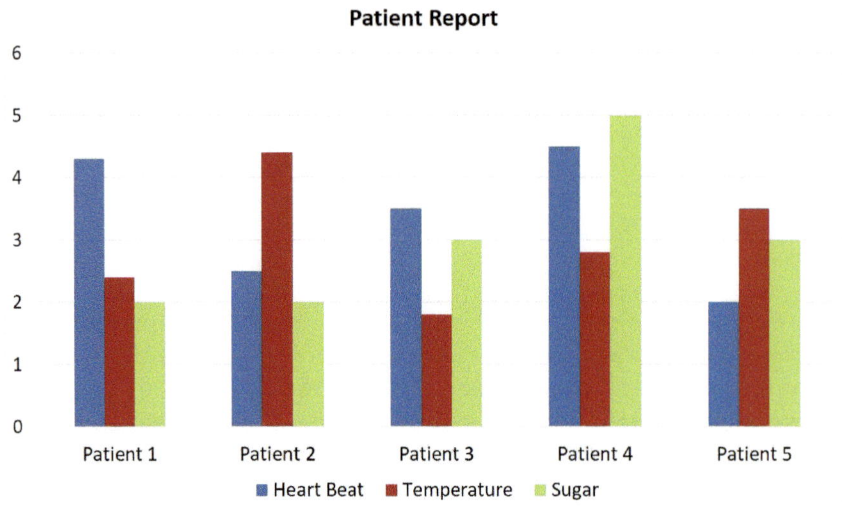

Fig. 10. Patient's report.

used to monitor the patient heartbeat, temperature, blood pressure, sugar level. The sensor measures the heartbeat, temperature, blood pressure and sugar. So this method is very useful for the patient life and also save the patient life and give new life to the particular patient. While the doctor on remotely we refer the patient report and analyze the patient body condition and give the suitable treatment and medicine to the patients. If the doctor is fetches any deviation to immediately help the patients. The doctor is continuously monitors the many number of patient. In physical treatment the doctor monitor the one patient at a time but The proposed BHMS allow the doctor to monitor the number of patient at that time and give the suggestion and treatment to the pharmacy supply chain is very important process in BHMS.

12. Conclusions

In this section conclude the chapter. In IoT SHMS have the lot of security issues such as the stored patient's data is easily attack or modified by third party. In Blockchain technology, the patient's data are stored by tamper proof. Due to the advantages of Blockchain is applied in HMS. So the stored patient's data cannot alter or modified by the third party. The proposed Blockchain Health Monitoring System (BHMS) is used to provide the security for IoT Secure Health Monitoring System (HMS). With the help of smart contract, BHMS is maintains the digital contract between the patients and HMS. In BHMS is used to maintain the security and support the remote services between the patients and doctors. In the future work, propose the consensus algorithm for secure transaction of Patient's Health Care Records (HCR).

References

[1] Devi, G. 2019. Security aware intelligent automation using internet of things (IOT). IJEE, 11(1): 62–67.
[2] Ashima, R., Haleem, A., Bahl, S., Javaid, M., Mahla, S.K. and Singh, S. 2021. Automation and manufacturing of smart materials in Additive Manufacturing technologies using Internet of Things towards the adoption of Industry 4.0. Materials Today: Proceedings, 45: 5081–5088.
[3] Muzafar, R., Singh, Y., Anand, P. and Sheikh, Z.A. 2022. Review on security of Internet of Things: Security requirements, threats, and proposed solutions. In Emerging Technologies for Computing, Communication and Smart Cities, pp. 747–756. Springer, Singapore.
[4] Navani, D., Jain, S. and Nehra, M.S. 2017, December. The internet of things (IoT): A study of architectural elements. In 2017 13th International Conference on Signal-Image Technology & Internet-Based Systems (SITIS), pp. 473–478. IEEE.
[5] Hurlburt, G. 2015. The internet of things... of all things. XRDS: Crossroads. The ACM Magazine for Students, 22(2): 22–26.
[6] Kumar, R. and Rajasekaran, M.P. 2016, January. An IoT based patient monitoring system using raspberry Pi. In 2016 International Conference on Computing Technologies and Intelligent Data Engineering (ICCTIDE'16), pp. 1–4. IEEE.
[7] Cao, G. and Liu, J. 2016, July. An IoT application: Health care system with android devices. In International Conference on Computational Science and Its Applications, pp. 563–571. Springer, Cham.
[8] Nidhya, R., Karthik, S. and Smilarubavathy, G. 2019. An end-to-end secure and energy-aware routing mechanism for IoT-based modern health care system. In Soft Computing and Signal Processing, pp. 379–388. Springer, Singapore.
[9] Adam, A.A.E. 2019. A survey of Internet of Things Technology and Projects for Health Care Services.
[10] Kajornkasirat, S., Chanapai, N. and Hnusuwan, B. 2018, April. Smart health monitoring system with IoT. In 2018 IEEE Symposium on Computer Applications & Industrial Electronics (ISCAIE), pp. 206–211. IEEE.
[11] Simić, M., Sladić, G. and Milosavljević, B. 2017. A case study IoT and blockchain powered healthcare. Proc. ICET, 1–4.
[12] Banka, S., Madan, I. and Saranya, S.S. 2018. Smart healthcare monitoring using IoT. International Journal of Applied Engineering Research, 13(15): 11984–11989.
[13] Onasanya, A. and Elshakankiri, M. 2021. Smart integrated IoT healthcare system for cancer care. Wireless Networks, 27(6): 4297–4312.
[14] Griggs, K.N., Ossipova, O., Kohlios, C.P., Baccarini, A.N., Howson, E.A. and Hayajneh, T. 2018. Healthcare blockchain system using smart contracts for secure automated remote patient monitoring. Journal of Medical Systems, 42(7): 1–7.
[15] Aqeel-ur-Rehman, S.U.R., Khan, I.U., Moiz, M. and Hasan, S. 2016. Security and privacy issues in IoT. International Journal of Communication Networks and Information Security (IJCNIS), 8(3): 147–157.
[16] Suo, H., Wan, J., Zou, C. and Liu, J. 2012, March. Security in the internet of things: A review. In 2012 International Conference on Computer Science and Electronics Engineering, 3: 648–651. IEEE.

[17] Agarkhed, J. and Patil, Y.D. 2017. A survey on Internet of Things towards issues and challenges. Journal of Innovation in Computer Science and Engineering, 7(1): 18–21.
[18] Wang, H. and Song, Y. 2018. Secure cloud-based EHR system using attribute-based cryptosystem and blockchain. Journal of Medical Systems, 42(8): 152.
[19] Griggs, K.N., Ossipova, O., Kohlios, C.P., Baccarini, A.N., Howson, E.A. and Hayajneh, T. 2018. Healthcare blockchain system using smart contracts for secure automated remote patient monitoring. Journal of Medical Systems, 42(7): 130.
[20] Xu, J., Xue, K., Li, S., Tian, H., Hong, J., Hong, P. and Yu, N. 2019. Healthchain: A blockchain-based privacy preserving scheme for large-scale health data. IEEE Internet of Things Journal, 6(5): 8770–8781.
[21] Dey, T., Jaiswal, S., Sunderkrishnan, S. and Katre, N. 2017, December. HealthSense: A medical use case of Internet of Things and blockchain. In 2017 International Conference on Intelligent Sustainable Systems (ICISS), pp. 486–491. IEEE.

Chapter 10

Blockchain Based E Voting System

Thomas Hanne,[1] *Deepak S,*[2,*] *Dhil Rohith B,*[2]
Sri Hari Nivas SP[2] and *Shobika ST*[2]

The following abstract will give a overview of electronic based voting system depend on block chain based technology. The only motive of this research will be research the current used methodology and proposed system of online based e voting system and related difficulties identification for future prevention and development. Online based voting system is a perfect replacement. Electronic voting solution with non-reputed, low secured applications. It also has a great capacity to minimize the organize cost and turn up with maximum no of voting. Even a minimum vulnerability can lead to a large number of vote manipulations in a voting system. Electronic voting system must be completely safe and protected, accurate, legible, convenient when used for traditional voting system. The another main motive of block chain based system is to eliminate the use of ballot papers and eliminate the voters from travelling from place to place for casting votes rather than casting from operating from wherever they are. Through various research and studies it is discovered that block chain based voting system is a perfect replacement

[1] Institute for Information syatems, University of Applied sciences and Arts, Olten, Switzerland.
[2] Bannari Amman Institute Of Technology.
Emails: Thomas.hanne@fhnw.ch; dhilrohith.cb20@bitsathy.ac.in; sriharinivas.cb20@bitsathy.ac.in; shobikast@bitsathy.ac.in
* Corresponding author: deepak.cb20@bitsathy.ac.in

and help to solve various problems in voting system. The major defined issue in this voting system is the speed of transaction and protection of privacy to a better develop of block based electronic voting system Privacy, transact speed must be developed.

1. Introduction

In every nation, the safety of a voting gadget is an effect of country wide safety. The current portfolio of portable computer security has studied for nearly a decade the potential of the current electoral system electronically, with the result of lowering the cost of having national elections, while entertaining and enhancing electoral security environments. Since a long time ago when democratically elected candidates were elected, the voting system was based on a pen and ballot paper. Anyone who has access to the system may be in danger of damaging the machine, which will affect all important votes cast in the above mentioned gadget. Engineers around the world have developed a new voting system, which offers some security while ensuring that the voting system must be accurate without fighting corruption.

Because of its simplicity, ease of use, and affordable price compared to known options, voting is widely used in many decisions. Almost all processes are currently in one place, authorized in the form of a special authority, controlled, measured, and monitored electronically, that is a clear voting problem in itself. However, digital voting systems have a single control over the entire voting system. Expanded society can be used as a modern voting system to win the most important mandate. The Block chain era provides a separate online voting booth or visual voting. The current block chain technology has been used to provide electronic voting systems primarily because of its advantages of securing to the end. The Block chain is an attractive opportunity for traditional virtual voting systems with capabilities that include segregation, non-discrimination, and defence protection. is widely used to maintain every board room and public vote.

A block chain, the first in a series of blocks, is a growing list of blocks mixed with crypto image connections. Each block contains a hash, timestamp, and activity information from the previous block. The block chain grows to be created to withstand realities. Voting is the latest block chain technology; in this area, researchers want to use the blessings and disclosures, confidentiality, and non-disclosure that can be important in voting processes. With the use of block chain digital voting packages, efforts and the use of block chain generation to protect and justify elections these days have found the entertainment of many people [7].

2. Block Chain Architechture

- **Node:** The nodes sincerely store, unfold and hold the block chain statistics, and thus it is able to be stated that a block chain exists on nodes. Nodes, consequently, are the framework of a block chain. Now, nodes may be any type of device, usually computers, laptops or servers.
- **Transaction:** it's far the block chain device's smallest structure block (NOTES and information), which block chain makes use of data.
- **Block:** it is a set of statistics structures used in technique transactions over the community dispensed to all the nodes.
- **Chain:** An order of blocks in a definite order.
- **Miners:** it is the nodes to make of transaction , upload that area into the block chain system.
- **Accord:** a set of instructions and agencies to carry out block chain tactics.

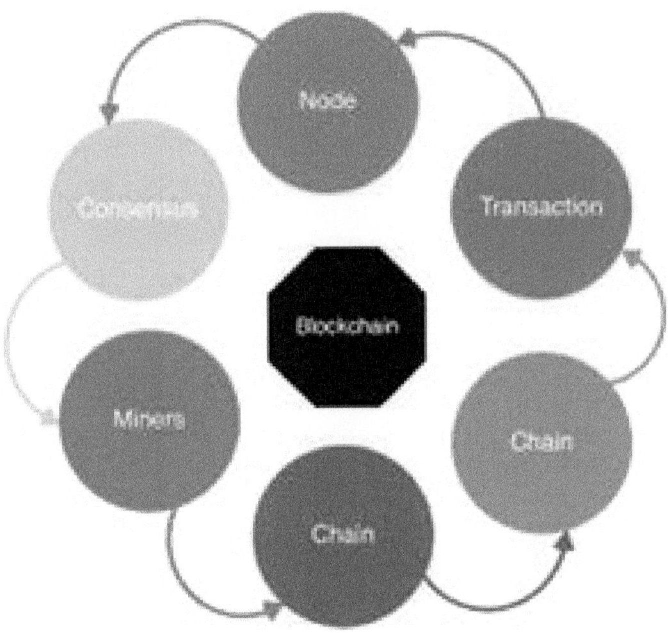

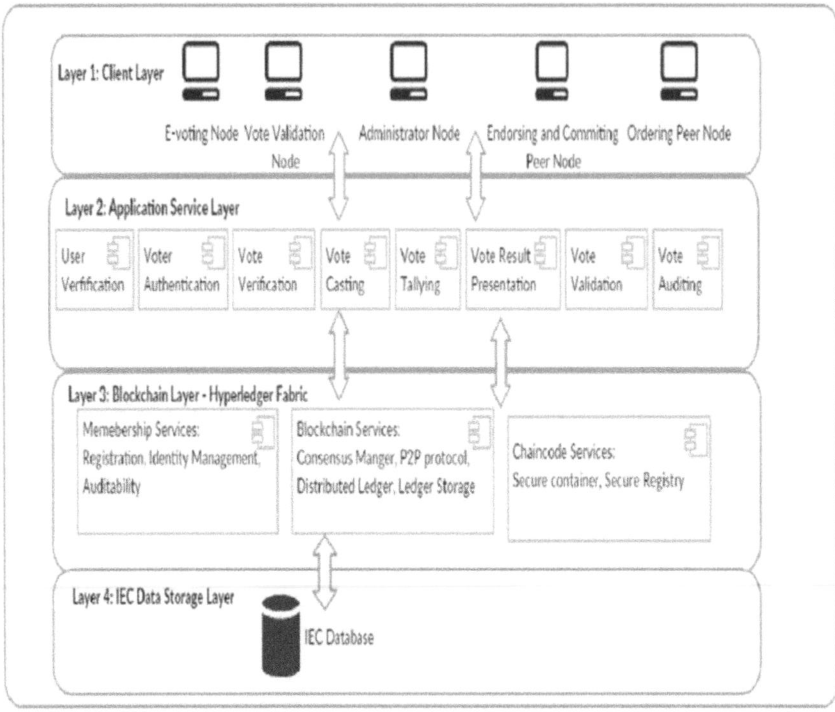

3. Electronic Voting Mechanisam in Blockchain

This segment offers a few history facts on digital vote casting strategies. Electronic balloting is a balloting method wherein votes are stored else counted electronic devices. Electronic voting is generally defined as voting to be supported by way of sure digital hardware and software program. Such regulators ought to be able to aid/put in force diverse capabilities from the electoral machine to vote storage. Kiosks, laptops and, greater lately, cellular gadgets in election workplaces are examples of all types of systems.

Electronic voting gadget ought to include voter registration, verification, vote casting and counting one of the areas in which the black chain may have a extensive impact is the. The risk stage is very excessive, e- balloting isn't viable choice the digital voting system has been hacked, inflicting some distance-attaining effects. Because the Black Chain network operates as an entire, centralized, open and consensual, it ensures that fraud is not theoretically feasible if the Black Chain based network design is not properly carried out.

As an end result, the particular traits of the block chain must be taken under consideration. There may be nothing inherent in black chain

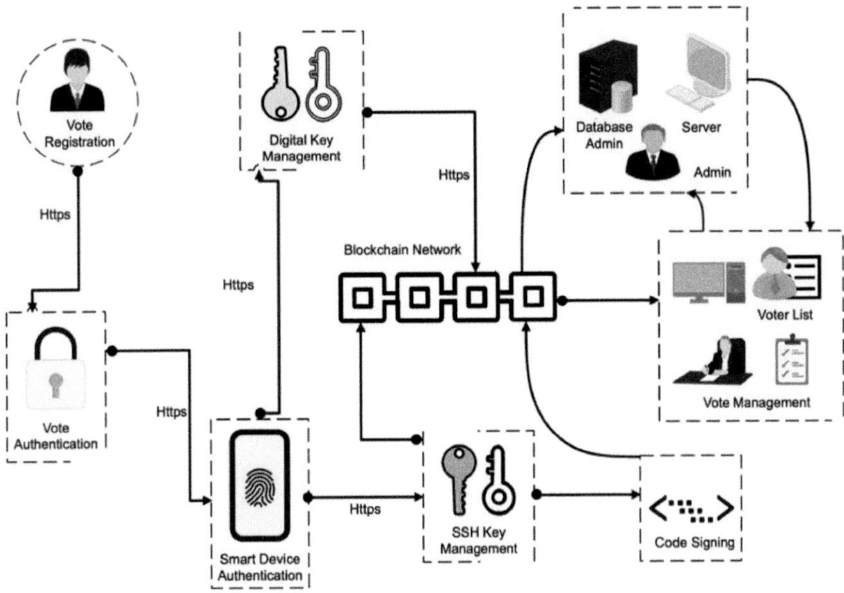

technology that forestalls it from being utilized in different varieties of cryptocurrency. The concept of the use of block chain technology by creating a tamper-resistant electronic or online voting community received momentum. Quit customers will now not see a significant distinction among a block chain based totally balloting device and a conventional electronic vote casting machine [2].

The block chain vote casting system is totally decentralized and fully open, but guarantees that voters are to be covered. This means that everybody can rely votes with the block chain electric balloting device without knowing who voted for whom.

4. Election Roles and Process

4.1 Election Rule

(i) As can be seen from the above considerations, elections in our view allow for the involvement of individuals or institutional bodies within the represented roles. Where more than one institution also individuals can be registered in the same category.

4.2 Election Admins

Control the existence cycle of an election. a couple of depended on institutions, organizations are enrolled with positions. The election

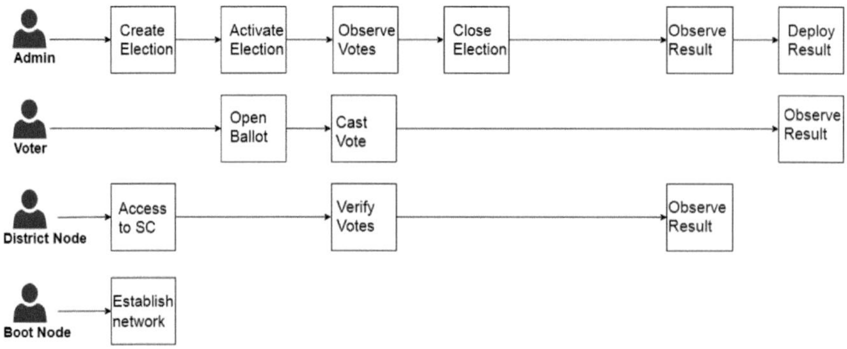

administrators specify the election type and create aforementioned election, configurator ballots, register voters, determine the life of the election and assign permissioned nodes.

Nodes

When election administrators make an election, each smart voting contract, representing each voting district, is placed in a series of blocks. When smart voting contracts are made, all parallel constituencies are given permission to communicate with their corresponding voting agreement. while the voter casts his or her vote in his or her smart agreement, the vote counts are verified by all corresponding regional nodes and each agreed vote is connected to a block chain while has been reached [3].

Voters

In the by-elections, citizens can vote for themselves, upload their ballots, vote their own ballots and confirm their vote after the election. Voters can be rewarded by voting with tokens when they cast their ballots in a nearby election, which can be challenging a major city.

Boot node

Each host, with access to the network, holds the starting point. The boot node helps the district nodes to find and connect. Boot nodes do not retain any block chain status and are used for static IP so regional nodes can detect their peers quickly.

A) Election Process

In our work, each election process is represented by smart contracts, which are initiated in the block chain by election administrators. Elections are defined as one smart contract for each polling district, so elections involve multiple smart contracts.

Each voter will be prompted with their respective polling district location, as defined in the voter registration phase, and the relevant location after the user authenticates himself or herself during the smart contract voter voting.

Create Elections

Election administrators create votes using the expansion plan. This geographical plan works with a smart election generation contract, in which the administrator defines a list of candidates and constituencies. This smart contract creates a set of smart voting contracts and applies them to the block chain, as well as a list of candidates, for each region, where each region is a candidate. Number on each voting for smart contract. Once elections have been created, each constituency is allowed to communicate with its smart voting contract.

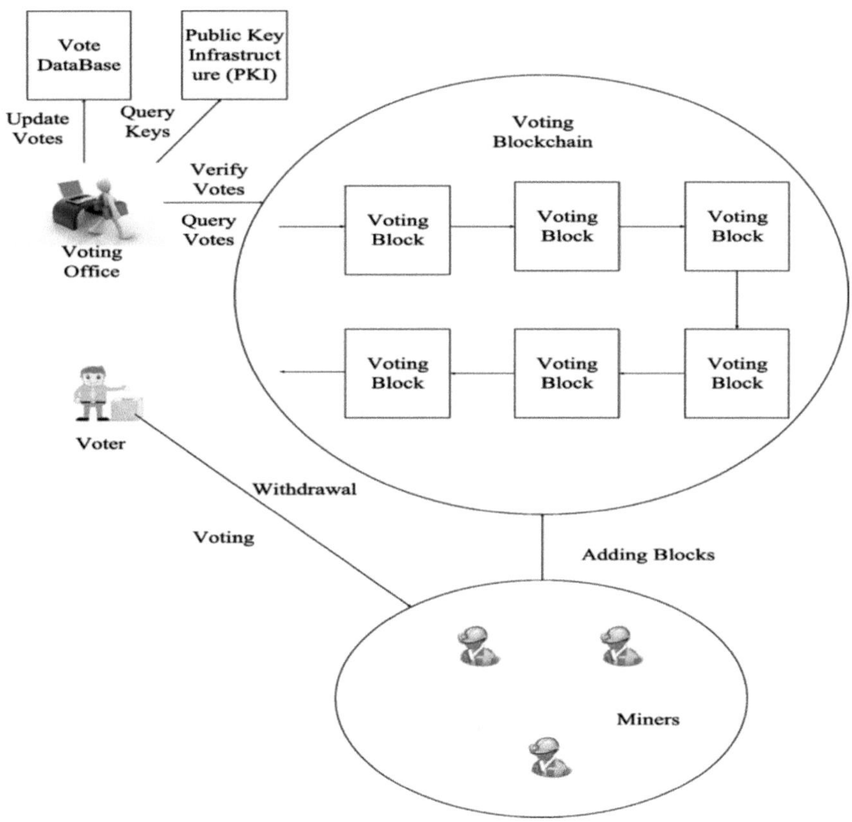

Voter Registration

The registration process is done by election officials. When creating an election, election administrators must provide a list of eligible voters. This requires a component that allows the government's identity verification service to securely verify and authorize the right people. By using these verification services, each eligible voter must have an electronic ID number and PIN, as well as information about the voter's residence. For each eligible voter, a corresponding fund will be created for that voter. The fund created for each voter should be different from each election in which the eligible voter participates and the NIZKP can be integrated to create such a fund so that the system does not know which fund fits each voter.

Vote Casting

While a man or a woman votes in a polling station, they deal with smart voting contract with the same voting as defined in any voter. This smart contract deals with the block chain with a corresponding regional node, which includes a vote in the block chain when agreement is reached between the general communities of corresponding regional nodes. The entire vote is kept in a block chain while everyone who casts their ballots gets their vote verified for verification purposes (see the "Confirmation of Vote" section). Everything done in the block chain has statistics about who will be voted on, and the location of the vote mentioned above. all votes are connected to the block chain by a smart balloon with a corresponding ballot, if at all if the corresponding regional nodes accept on the check of the ballot statistics.

When a voter casts his or her vote, the burden on his or her fund is reduced by 1, so he or she does not allow them to vote more quickly than each election. As can be seen in the above, singles transactions in a series of general public blocks include job ownership, the block where the transaction will be located, the age of the transaction, the wallet that sent the transaction and who received it, all payments. Sent along with payment for transaction. What is being done on our proposed device does not require all these statistics, only one function has job id statistics, the block where the task is placed, where the smart contract function is converted into a post, in this case the N1SC shows. That the vote be cast in the N1 constituency in the end the price is vote included in this purchase be a birthday party vote. The age of singleness does not include the protection of every citizen from the onslaught of time.

Tallying Results

Election counting is done in an instant on smart contracts. Each smart ballot contract creates its own corresponding space its storage. Once the election is over, the final result of each smart contract is published [1].

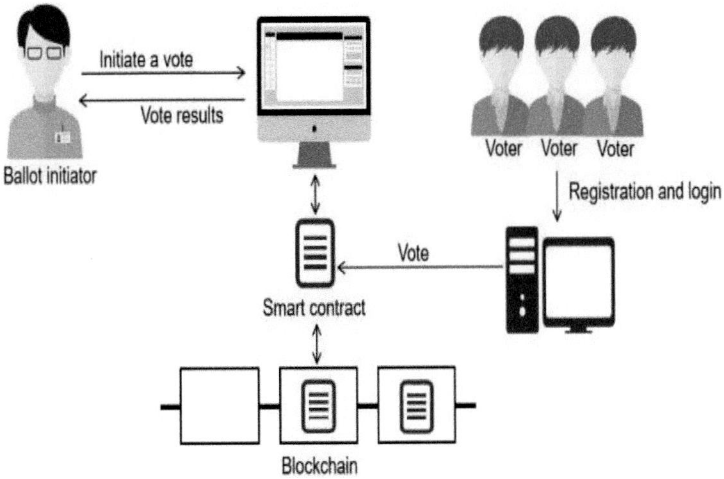

Verifying vote

As noted earlier, every male or female voter receives ownership of his or her vote. each voter can visit a government official and present their work ID after verifying their use of digital identity and the corresponding PIN. Officials, using the regional node access to the Block chain, use block chain explorer to find work with the corresponding action id in the Block chain. The voter can therefore see his or her vote on the block chain, which ensures that it is converted and counted correctly.

4.3 *Evaluation and Implementation*

4.3.1 Evaluation

The main reason of the test was to assess the effectiveness of the device by using looking at e-balloting the machine requirements supplied in segment 2 and the identification of any issues on this regard utility inside the actual global. The check consisted of several steps namely practice more than one transactions, transaction confirmation, block chain mining transactions, demonstration of changes made to the general public listing on all nodes in the community and machine usability.

The test changed into done at once on Multi chain by way of to begin with developing belongings. The result of this proven by means of it. We might also confer with these assets as votes [4]. As Multi chain automatically suits cryptocurrency, so we wrote our APIs to design it inside the context of the vote. To do transaction in Multi chain, we've got identified the cope with and stability in the node deal with of Multi chain from wherein the products (vote) can be sent. Even as the products had been being despatched to the address, a piece hash turned into produced sporting the transfer of votes. The balance of the reception place is extended by way of one vote (belongings). As our custom asset design API is designed to deal with he will have handiest one vote (assets), therefore, it will not be feasible for a voter to cast a majority vote until the node unearths it in some other address handiest allowed in candidate mode [8].

4.3.2 Implementation

The implementation of the proposed plan was done within a controlled area with a web-based application designed to function as a pre-existing system that allows users to interact a easy way out. This application is running in Java EE within the Net beans are Indigenous Glass fish server used to host the application. Glass fish server-side server managing application EJBs and data source. The application uses MySQL as a back end Application website also contains administrator data manually entered as voter information, regional details and information about the various.

Political parties contesting the election. A screenshot of the application showing the administrative function to view the list of eligible voters is displayed in Fig. In addition to manual installation, the app also supports data import using MS Excel spreadsheet bulk import sheets by viewing the data size in real-world voting conditions. Use it Multi chain as a block chain platform to build a private block chain for this application used recording voting activities. This choice is influenced by the ease of use provided by this forum it was therefore easily integrated.

4.4 Security Requirements for Voting System

(1) Anonymity: Throughout the voting process, the number of voters must be protected from external information. Any relationship between registered voters and voter identity in the event of an election will not be known.

(2) Voter Privacy: After voting, no person should be in the party to attach voter identity and vote. pc secret is a soft form of confidentiality, due to the fact that voting is always hidden for a long time as long as the current price continues to change with the power of PC and new technologies.

(3) Stability and Integrity: The condition that a large enough party of voters or representatives will not interfere with the election. It ensures that voters can avoid them without problems or encourages others to cast their own ballots. Corruption of citizens is illegal in preventing the outcome of an election by arguing that a number of different members no longer complete their role properly.

(4) Readability and Accuracy: Accuracy, also referred to as accuracy, requires that the results match the declaration of the election. It is a way that no one can change the vote of other citizens, that the final figures include all valid votes, and that there are no exact numbers invalid votes.

(5) Democracy/Unity: The democratic gadget is defined where only eligible person can vote, and a simple single vote can be valid for every registered person any other factor that no one else would be able to repeat vote.

4.5 Mining of Voting Blocks

All votes inside the block chain crypto graphically block by means of block by block. Many comfy hash algorithms may be used to resolve the trouble of encrypting a message within the modern-day block to generate a message alert, which includes SHA-256.

The brand new block is generated by way of customers from the P2P community. The brand new era of block is based totally on the PoW set of rules [5]. Once a new vote is submitted and verified, the miner generates a brand new block with balloting data and distributes new data to network. If fresh data have a simultaneous stamp, a block with a high signature cost is selected over the others.

4.6 Benefits and Disadvantages of the System

In a democratic vote casting method, safety and integrity are paramount. To reap this, it is necessary to make sure an extended listing of requirements for instance:

1. A voter can handiest vote as soon as.
2. Their anonymity and vote casting secret have to be maintained.
3. There have to be no way to verify that a person has voted (in nations where voting is obligatory) and/or the only who voted for them.
4. Balloting must be easy and reachable.
5. No voter or candidate can misinform the effects.
6. There's a need for openness and auditing within the voting gadget.

One of the problems with e-balloting is that it's miles nearly impossible to satisfy all the necessities. Apprehend vote casting electronically like all vote casting gadget saved on a particular hardware or software. We have some outstanding and smart technology that we can use in those systems, however although, they almost continually want to be traded. To verify one (or greater) necessities, we need to discontinue one or more of the others. With Block chain, it'll be no different. However no matter this, it's far nonetheless promising.

Block chain fraud does not take place visually if performed correctly. Votes can be transferred as an asset to block chain, and altering or deleting votes will now not be feasible. The vote casting method will no longer depend upon the organization, institution, or authorities, that is, the relevant authorities.

In fact, that is the principle in the back of Decentralized self-reliant companies (DAOs) where companies are governed through the equal principle in their regulations, operations, and selections are saved in Block chain. Those organizations function independently and as everything is carried out with Block chain, the regulations follow equally to all contributors and the whole thing is obvious, making corruption even greater hard. This shape permits DAOs to be self-prepared and non-compliant. Any other common use of DAOs, due to their symbols, vote casting and democratic decision-making, because it allows for open, obvious, and segregated participation.

Votes will be stored publicly on Block chain and disbursed across community nodes instead of saved on crucial servers. consequently, every person can take a look at the votes and comply with the balloting method, whether or not voters or candidates, and (in all likelihood) will now not recognize who they voted for because they do no longer use usernames or something like that.

You may have noticed that "perhaps" wrapped in brackets in the sentence above. That is due to the fact it's far possible to recognize who voted for you in a positive manner by using "tracing" or checking someone to find out the cope with in their node. Additionally, you should be very cautious about Block chain scalability issues. Because of its nature, that requires calculation power and consequently calls for time, due to the fact Block chain is sluggish. The technique of creating new blocks is time eating. Which means that for huge votes, the Block chain-primarily based balloting device could have issues. Now not to mention the amount of power that it turned into going to devour [6].

Now that we recognise a number of the professionals and cons of Block chain, permits check one of the approaches to use Block chain in digital voting. The city of Moscow has lately attempted this idea, and this check is what we are able to talk about! Destiny paintings.

4.7 Social Impact Use Cases

Fair and free democratic elections are main features of a country. As the humans become more digital and technology is wide spread, e voting and bloc chain technology has more scope in the field of politics. And make voting most access.

To every people and improve integrity of election all over the world. Because e voting made of block chain technology is more appealing.

4.7.1 Future Work

Digital voting has been used in numerous ways for the reason that Seventies with tremendous benefits over paper-based structures along with multiplied performance and decreased errors. With an unusual increase within the use of block chain generation, many attempts were made to test the feasibility of the usage of it block chain to facilitate an powerful voting solution by way of e. This paper gives one such attempt finds block chain benefits including crypto picture foundations and transparency for the gain of powerful solution for voting by way of e. The proposed method used with Multi chain and an in-intensity examination of the technique highlights its effectiveness on the subject of fundamental gains.

The necessities for an digital balloting gadget.in the continuation of this paintings, we're focused on improving the resistance of block chain era [9].

The trouble of 'twin use' in an effort to be translated as 'twin vote casting' into digital balloting systems. Despite the fact that Block chain era achieves extremely good fulfilment in accomplishing an easy transition in buying and selling but a success demonstrations of such events had been received which inspire us to research constantly. Up to now, we consider the operating version to set up a dependable foundation for balloting through e systems may be crucial to reaching a digital balloting device with give up-to-give up verification. The project of accomplishing this keeps within the shape of an additional layer of provenance to assist the present block chain based totally infrastructure.

(1) Scalble Processing Overhelmes

With a small range of customers, the block chain works well. However, whilst the network used for larger options, the wide variety of customer's increases, main to better costs and well timed consumption of processed ingredients. Increase problems increase with a growing wide variety

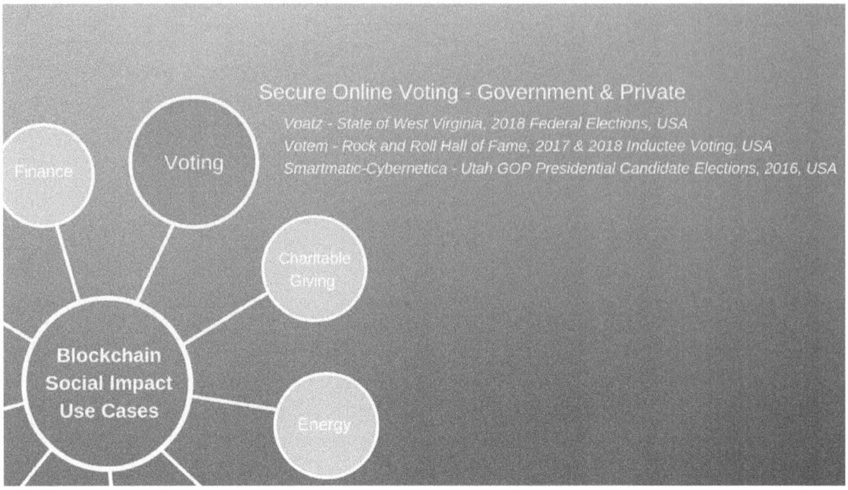

of nodes within the block chain network. Inside the case of elections, I systemic degeneration is already an critical trouble. digital balloting integration may be completed it also contributes to block chain-based totally machine scores specifies the special metrics or properties available throughout the block chain framework and presents comparisons evaluation of other block chain-primarily based structures along with Bitcoin, Ethereum, Hyper ledger fabric, Lite coin, Ripple, Doge coin, Peer coin, and so on. Any other way to improve block chain score be a metaphor, called sharing. In a well-known block chain network, sports and blocks are demonstrated by means of all taking part nodes. to present excessive strength consistent with the information, the facts ought to be horizontally divided into segments, each called episode.

(2) Immatureness

Block chain is a flexible technology that symbolizes complete flexibility in a divided network. It has the energy to transform organizations in keeping with approach, structure, techniques, and cultures. The modern-day block chain implementation is non-existen terrors. Technology is vain right now, and there may be little public or professional information of it, making it impossible to test their future skills. Everything in technology troubles in block chain capture are regularly because of technical immaturity [10].

(3) Accaptableness

At the same time as the block chain is at the vanguard of turning in accuracy and security, human self-belief once and for all trust is an essential element in powerful block chain vote casting. Problem of block chain can make it tough for humans to accept digital voting based totally block chain, and can be a giant impediment to in the long run accepting block chain-based electronic balloting by using preferred public reputation. A first rate advertising marketing campaign needed for this motive in an effort to to elevate attention approximately the advantages of block chain voting systems, so to talk make it easier for them to embody this new era.

5. Conclusion

We've delivered a digital block chain-primarily based digital voting machine that makes use of clever contracts to allow secure and cost-effective elections even as making sure voter privateers. We've got described device configuration, design, security analysis of gadgets. compared with previous paintings, we've got shown that block chain era offers with new possibility for democratic nations to broaden from a traditional election device, to a better value-andan extended-time period electoral system, while improving the safety measures of present day gadget and offering new opportunities for transparency. The usage of Ethereum's non-public block chain, it's far viable to ship loads of sales in keeping with 2d to block chain, uses all aspects of a smart agreement to simplify the burden at the block chain. In international locations of big size, additional measures must be taken to decrease immoderate consumption of in line with 2d interest, as an instance figure-child creation which reduces the wide variety of activities saved within the block chain by 1: a hundred without compromising community safety. Our electoral machine allows individual citizens to vote in for the constituency of their desire even as ensuring that each individuals vote is counted from the correct, which may additionally increase the quantity of citizens.

References

[1] Tan, W., Zhu, H., Tan, J., Zhao, Y., Da Xu, L. and Guo, K. A novel service level agreement model using blockchain and smart contract for cloud manufacturing in industry 4.0. Enterp. Inf.Syst. 210.

[2] Steve Ellis, Ari Juels and Sergey Nazarov. 2017. ChainLink: A Decentralized Oracle Network Available at: https://link.smartcontract.com/whitepaper.

[3] Hakak, S., Khan, W.Z., Gilkar, G.A., Imran, M. and Guizani, N. 2020. Securing smart cities through blockchain technology: Architecture, requirements, and challenges. IEEE Netw., 34: 8–14.

[4] Konstantinos Chalkias. 2017. Demonstrate how Zero-Knowledge Proofs work without using maths Available at: https://www.linkedin.com/pulse/demonstrate-how-zero-knowledge-proofs-work-without-using-chalkias.

[5] Alin Tomescu and Srinivas Devadas. 2017. Catena: Effificient Nonequivocation via Bitcoin Available at: https://people.csail.mit.edu/ alinush/papers/catena-sp2017.pdf.

[6] Feng Hao and Piotr Zielinski. A 2-Round Anonymous Veto Protocol Available at: http:// homepages.cs.ncl.ac.uk/feng.hao/fifiles/av_net.pdf.

[7] Yaga, D., Mell, P., Roby, N. and Scarfone, K. 2019. Blockchain technology overview. arXiv 2019, arXiv:1906.11078.
[8] Ometov, A., Bardinova, Y., Afanasyeva, A., Masek, P., Zhidanov, K., Vanurin, S., Sayfullin, M., Shubina, V., Komarov, M. and Bezzateev, S. 2020. An overview on blockchain for smartphones: State-of-the-art, consensus, implementation, challenges and future trends. IEEE Access, 8: 103994–104015.
[9] Jelurida Jelurida. 2017. Available at: https://www.jelurida.com/sites/ default/fifiles/ JeluridaWhitepaper.pdf.
[10] Agora. 2017. Agora: Bringing our voting systems into the 21st centuryAvailable at: https://agora.vote/Agora_Whitepaper_v0.1.pdf.

Chapter 11

A Secure and Efficient Student Performance Prediction System Based on Blockchain and Data Mining Technologies

Parvathy S Kurup[1,*] and *SV Annlin Jeba*[2]

In the Educational Data Mining scenario, the academic institutions want to securely collect and store student data records for knowledge discovery in educational databases. Blockchain technology is a distributed technology that can securely store and exchange digital information. It can be used effectively for the safe storage of educational data. By securely storing personal and educational student information on a distributed database system, we can prevent data fraud and data misuse. Then by analyzing the data educational institutions can classify students based on their learning outcomes and can reduce the rate of failure and dropout. The ability to predict students who fail academically can be used effectively to advise the student and to guide academics to make the necessary interventions in everyday subject matters. The

[1] PG Scholar Department of Computer Science and Engineering, Sree Buddha College of Engineering, Pattoor.

[2] Associate Professor and H.o.D, Department of Computer Science and Engineering, Sree Buddha College of Engineering, Pattoor.

* Corresponding author: parvathykurups@gmail.com

proposed system introduces a two-phase model that combines Blockchain technology with predictive statistics. Blockchain technology is used here to securely store and share student data records. These records are provided in the Predictive analytics model. The predictive analytics model incorporates statistical analysis methods and classification methods. The statistical analysis is based on Multiple Linear Regression which is used to identify many influential factors that affect a student's final grade. Student moral, physical, and psychological qualifications are provided with a predictive analytics module to identify students at high risk of failure. Here we have used four distinct classifiers to predict future student performance. Among these, we have identified the most effective classifier - Artificial Neural Network (ANN) which performs better than other classifiers with an accuracy rate of 93%. By using the proposed system, we can improve the quality and security of education by getting better results.

1. Introduction

The education sector embraces new technologies to improve student academic performance and learning outcomes. Technologies such as blockchain and Educational Data Mining can be used effectively in the field of education to securely store and analyze student data records. Blockchain technology has gained widespread acceptance due to the growth of digital finance. The Blockchain mechanism is a distributed, distributed technology that ensures data security. It can be considered in banking, educational sector, management of supply chain, and protection of copyrights. It is similar to log data with all activity listed in its order of occurrence. Blockchain contains a long series of interconnected dataobjects which are distributed across various devices. This study aims to present the use of Predictive Analytics and Blockchain technology in education field. The use of Blockchain provides an effective platform for securely keeping student data records to avoid data fraud and misuse. Today all higher education institutions face challenges in improving student achievement levels. To improve students performance the institutions must first predict the failure rates of students and then give them proper assistance to overcome the barriers to study. This can prevent failure rates and dropping out of school [1].

Each higher education institution should implement a student performance management system based on the legitimate and credible features of students who select people who may be successful in their studies. Each university must use a system to predict the future grades of students in early stages of study itself. Institutions of higher learning face many challenges when analyzing their huge academic data to capture information about student performance. This situation arises as a result of the use of traditional statistical methods instead of using the new technology like Educational Data Mining (EDM), which analyses student data and predict student learning outcomes.

Educational Data Mining is a process of finding out relevant information from major educational repositories and using this feature to predict future student marks. By using this learning outcomes can be further improved and can predict vulnerable students in advance. EDM focuses on mining knowledge from the context of education to solve educational problems. EDM's sub-area is Learning Analytics [2]. It analyzes student data to identify student activities to provide appropriate recommendations for the development of student knowledge. Predictive Analytics is part of Learning Analytics and provides analytical support to the education system. Predictive Analytics is based on predicting student performance based on predictable factors and methods [2]. Each university should aim for a performance appraisal system based on a legal and credible issue that separates the people who will be successful in their studies.

The proposed program supports educational institutions to classify their students so that they can provide the necessary assistance to achieve better outcomes. It predicts applicant academic performance through student admission records that include personal and educational features. In particular, the proposed approach addresses the following objectives:

(1) Integrating Blockchain technology with Predictive Analytics to develop a secure student performance prediction system.
(2) Identify key factors that affect student performance.
(3) Divide students based on academic and non-academic aspects.
(4) Finding the most accurate data prediction method.

The paper is structured as follows Section 2 contains Related Works, Section 3 explains Methodology, Section 4 contains Results and Discussions, and Section 5 presents Conclusions.

2. Related Works

The education sector is embracing new technologies for the betterment of the quality of education. Blockchain and machine learning are the two technologies that have recently been used by the education sector for further development. This section describes the literature review on various blockchains and Machine Learning technologies used in educational sector.

The application of block chain technology is in business, digital crypto currency and it is now used in the upliftment of educational sector [3]. It is a distributed record of digital events, a distributed consensus approach to agreeing that a new block is legal, smart automated contracts, and the data structure associated with each block [4]. The study involved a preliminary evaluation of a private blockchain for academic records. Blockchain technology allows for the creation of a distributed digital event record in a distributed environment where data and related activities are independent of any third parties. It highlights existing problems related to educational institutions and to find appropriate blockchain features that they can solve. In another study, the authors used a systematic review of the literature to find and extract relevant information from short-listed subjects [5]. It also describes the problems that exist in the three physical, digital and financial spheres. Research findings indicate that the risk of fraud, difficulty verifying, and sharing of records between institutions are major problems educational institutions face. Another piece of literature proposed a Technology Adoption Model (TAM) [6] with additional flexibility such as visual security, privacy and trust. Related literatures gives a clear-cut explanation of the effectiveness of blockchain technology in secure storage and distribution of digital data.

The quality and reputation of all higher education institutions depend on the consistent learning outcomes of their students. Educational data mining is used to analyze data and to predict student learning outcomes. To predict student performance, the main factors to consider are attributes and predictive methods. Many studies find a correlation between points of admission process and graduation marks [2]. The previous year grade points alone are not enough to predict the outcome of a student's graduation but may be a useful guide. One of the studies has shown that a student's racial background means nothing in predicting their performance [2]. Long-term employment and family size have impaired academic performance [7]. The personal background has played a very important role in predicting student performance [8]. A Study-based research has used a combination of an online learning environment and an online practice site with classroom teaching to calculate performance prediction based on reading levels in question and homework [9]. Giving prior attention to subject-specific data specific to subjects improves grade prediction accuracy [10]. Previous courses may provide students with information on future courses so that previous course marks can be used to predict marks in future studies [11]. One study suggests that student performance is an indicator of the quality of educational institutions [12]. The total CGPA score alone cannot be considered a parameter to assess student growth and gender does not reflect student performance [12]. Demographic information, socio-economic status, high school background, intermediate grade enrollment, high school final grade, etc., are considered for predictions in another study [13]. They proposed a segmentation framework to classify at-risk students. Many factors can affect student performance, one of the research projects shows that history marks, student activities in academia, student mathematical knowledge, and student are some among them. Related studies emphasize that predicting student performance using academic data mining reduces the failure rate by predicting students at risk and reducing the tendency to drop out of school. A few key factors that affect student performance in a single subject are the CGPA of the previous study and a few socio-economic factors as well as the individual characteristics of the student. In all books, pre-admission records or personal characteristics are used for prediction, but your combination of both is not used. Guessing models developed for many types of literature are based on common methods of retrospective and subdivision. Many of them are based on a supervised learning method. The most common mathematical and machine learning algorithms used to predict are Multiple linear Regression(MLR), Principal Component Analysis (PCA), Linear Discriminant Analysis (LDA), Multi-Layer Perceptron (MLP), Artificial Network Neural (ANN), Support Vector Machine (SVM), Decision Trees, Naïve Bayes, etc. The ANN performed much better than most other classification strategies in most studies [7, 14]. The combination of PCA and MLR improves the accuracy of the prediction model [9]. Predicting continuous values or features that include two or more regression methods can improve the accuracy of the forecast. The mixed retrospective model used in the study [15] showed significant improvements compared to single-core models and demonstrated the usefulness of the proposed method of identifying multiple factors that affect student performance. Most studies have used small data sets to predict [11, 13, 14].

Related literatures provide a clear definition of the various data sets, strategies, methods, and models for predictive analytics. They discussed the possibilities of training and modeling data sets and their potential for creating a more accurate prediction model. Assess the feasibility of identifying key indicators that have a significant impact on prediction outcomes. All the studies were very focused on improving the quality of education provided to students. The main focus was on improving the quality and reputation of universities nationally and internationally. The growth of student knowledge considered and the impact of early education on student performance has been identified. Some of the subjects used a small dataset with fewer attributes and some used a real-time database that has both student and social attributes. Supervised, unsupervised, and unsupervised learning methods are used for predictive analysis. Each method has its drawback and is directly related to the nature of the dataset used in the study. Some of the studies have explicitly used WEKA to predict. Therefore, we have performed a comparative analysis of results obtained in both WEKA and Python. The results showed that WEKA software is trustworthy.

3. Methodology

Blockchain technology is a very transparent, distributed, and flexible ledger that can be used as a secure place to store and share student data records. Student data records contain student academic knowledge, socio-economic behaviour, and health-related information. Some of them are very sensitive to being managed and require high confidentiality. In this context, the blockchain can be used as a secure archive for keeping student data records to prevent data fraud. In the proposed program integration of Blockchain and machine learning technology is used to enhance student academic performance. A standard, accredited, decentralized database that can keep all student records is a must factor of all educational institutions. These data records must be kept secure so that no fraudulent data or unauthorized access can occur. This secure storage method thus ensures system integrity. The proposed approach introduces a two-phase model that combines blockchain technology and Predictive analytics blockchain technology used here to securely store and share student data records. These records are provided in the Predictive analytics model. The predictive analytics model incorporates mathematical analysis and classification methods. The mathematical analysis is based on Multiple Linear Regression which is used to identify many influential factors that affect a student's final grade. Student moral, physical, and psychological qualifications are provided for predictable models to identify students at high risk of failure. Here we have used four distinctions to predict future student performance. The proposed program consists of three stages including data-collection, data-storage, and data-analysis. Figure 1 depicts the proposed system architecture.

3.1 Data Collection

The proposed system uses a record of 524 students studying in various schools during the 2018–2020 period. The scheme uses a standard data

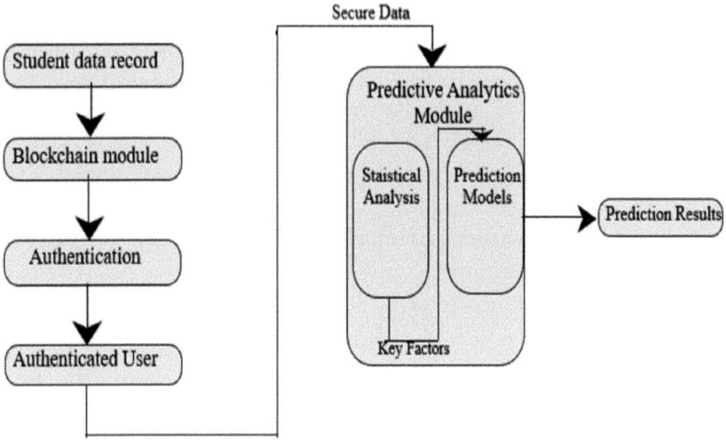

Fig. 1. The approach for a secure and efficient student performance prediction system.

analysis method which can be applied to any academic institution. The data is collected and processed first using Microsoft Excel. Preliminary processing of data involves the external acquisition, removal of insignificant values, and aggregation of data. The dataset was modeled using binary classification and regression tasks. It is a multivariate dataset that includes 524 instances and 30 attributes. All data sets are organized into Microsoft Excel. Attributes included previous grades of students, demographics, socioeconomics, and school-related factors, and were compiled using reports of students and by questionnaires. The Excel file is then converted to the Comma Separated File required by the WEKA tool and Python. WEKA is based on java and it is an open-source tool which is widely used in EDM research field. A data pre-processing is done on the dataset to eliminate irrelevant attributes, identified resources, deleted invalid values, redundant values. We have built target (class) variables based on the parameter of real-time grade last year. It has two categories based on the following scope: Pass (≥ 9) and failure (< 9). The same data set is also provided with Python IDE for prediction.

3.2 Data Storage

The attributes in a student data record include social, psychological, behavioral, economic, and student characteristics. These include parental education, parental status, family support, economic background, preschool records, school infrastructure, school support for poor children, government support, alcohol consumption, unemployment, health record, outing with friends, social networking, status teacher's attitude, etc. This is very sensitive information, so the data used for processing should be protected. To protect the data, we can use the blockchain method, which is a site-divided site, which contains a series of data blocks that form a chain. A data chain consists of multiple copy of the same data and incase a chain is broken, other chains of the same data are available. Therefore, we use a blockchain method for student data records.

Each student enrollment number can be used as a unique identifier to add student details to the block chain. This unique ID can help track transactions between unknown entities in any digital transaction. The features of the blockchain are:

1) Ledger: A blockchain that incorporates each action based on its emergence. Retail data is not modified and is distributed to all nodes.
2) A series of blocks: Contains a series of interconnected blocks where every block contains the value of the previous block constitutes the chain.
3) Confidentiality: Promises confidentiality of data without disclosure to an unauthorized person.
4) Transparency: Using the algorithm for agreeing order and execution of boxes is guaranteed.
5) Cryptology: Provides encrypted functions based on generating fixed hash values in data blocks.

The above features explains the effectiveness of blockchain technology to be used as a secure data storage system.

3.3 *Data Analysis*

To date, the security features of the system are being discussed. The third component of the system is a predictive analytical model based on Machine Learning. The blockchain method provides authorized data to the predictive analytics model to make accurate and reliable predictions. Here secure student data is given directly into the predictive analytical model which includes mathematical analysis and prediction models. Predictions are used to classify students and to give vulnerable students the right direction to overcome failure. The most important factors for students influencing their learning outcomes are determined and use these factors to divide them into two groups to improve their academic performance.

The learning method used here is supervised learning and models are developed using classification and regression. To analyze and to visualize data Exploratory Data Analysis (EDA) is used. Regression is used to understand the relationship between attributes. It predict output labels or responses to the provided input data. The purpose is to predict values that are as close to the actual output as our model can. Model accuracy is done by checking the error number such as Root Mean Squared Error (RMSE) and Mean Squared Error (MSE). Classification is a supervised learning approach method that uses an algorithm to classify labeled results, where the output has defined labels or different values. In this scheme, we want to separate those students who are likely to fail the exams. Here we used eager learning classification algorithms like Support vector machines (SVM), Random Forest, Artificial Neural Networks, Naïve Bayes, etc.

Figure 2 represents the block diagram of the Predictive Analytics module which can be divided into mainly five parts. They are:

1. Development Environment
2. Data Analysis
3. Statistical Analysis
4. Predictive Analytics

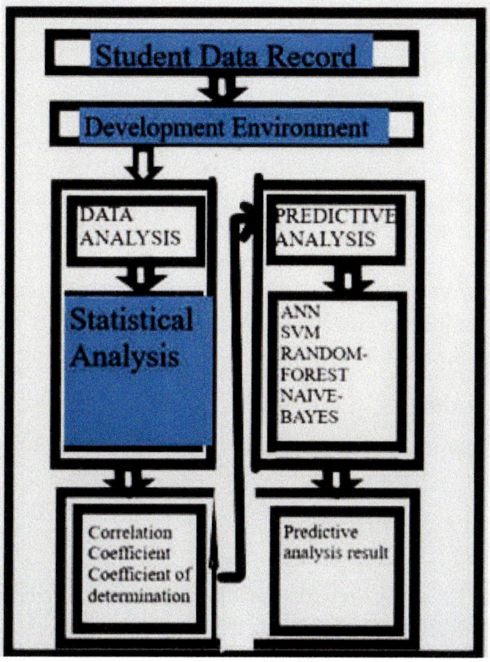

Fig. 2. A detailed overview of predictive analytics module.

Development Environment

Here two development environments are used to evaluate the regression as well as classifier accuracy. The first one is the Waikato Environment for Knowledge Analysis (WEKA), an open-source tool for doing machine learning and data mining applications. It contains built-in storage that includes almost all machine learning algorithms, so we do not want to program the code explicitly. The CSV file is imported into WEKA in the form of Attribute Relation File Format(.arff). An attribute subset selection method is used here to select the most relevant attributes. Both Statistical analysis and Predictive Analytics were performed using WEKA.

The second development environment used is Python3. Jupyter notebook is used as the IDE for programming. Libraries like NumPy, Pandas, Matplotlib, Seaborn, and SK learn. NumPy and pandas are used to manipulate and analyze data. Matplotlib and Seaborn are used to view data. SK Learn or scikit learn contains all the machine learning algorithms. Each model was explicitly programmed and evaluated the result.

Data Analysis

Both WEKA and Python3 Exploratory Data Analysis (EDA) were used to visualize the target distribution of variables and the visualization of variables. For that histogram, bar charts, and scatter pieces are used.

Statistical Analysis Linear Regression

Data in the development environment is first used in the data analysis function to identify relevant features. Multiple Linear Regression model is used to find the most relevant feature using the correlation coefficient analysis, i.e., finding the relationship between multiple factors (independent variables) and dependent variables (academic grades). The result revealed that there is a strong positive correlation between variables such as previous year's grades, father and mother education, economic background, teacher attitude, etc. Negatively Related Qualities Alcohol Abuse, Outings, Social Networking, Student Attitude, etc.

Predictive Analysis

Predictive analysis part consists of four prediction models explained as follows.

Artificial Neural Network (ANN)

ANN imitates human intelligence to find solutions to complex real-world problem. Its architecture contains a set of neurons which receive a activation function and analyzing the weight it produce some outputs. Variable-variable interaction can be analyzed by using ANN and it can learn from small sample set . Topology used here is Multi layer Perception (MLP) to develop ANN because the dataset is not so large.

Random Forest

Decision Tree contains root nodes, intermediate nodes (leaf nodes). A set of test data is used in the middle and after some processing the output labels are predicted and the result is stored in leaf nodes. As in Ref. 9 and 10 here, the central area represents the model element, and its branches represent the possible values for that event [8, 9]. We have used this process because Random Forests are very easy to use and easy to manage, results are predictable directly.The rules set by random forest can be easily understood and well suited for predictions.

Support Vector Machine (SVM)

The supporting vector machine is based on supporting vectors, which are data points that are very close to the main gene line. This separation process creates a hyperplane which is a subset of vector spaces of a limited size similar to straight lines in planes and planes in 3D space. The hyperplane classifies data points based on their similarities. If the datapoints are too far away from the hyperplane, the SVM standard error decreases significantly. It models small datasets very efficiently. Its computing efficiency is very faster.

Naive Bayes

It is an easy-to-use method relies on Bayes theory and probability distribution. Provides opportunities for each item in each class that is possible. It is simple, efficient to solve real world problems, faster computing efficiencies, and used many times in other literatures.

WEKA: Data with relevant features are fed into each of the Machine Learning models and the output (prediction) of each model is obtained. ANN gave the highest accuracy, 94% and Naïve Bayes gave the least 90%. Stratified cross-validation is used to evaluate the models accuracy.

Python: The Data set is first splitted and fed into each of the classifiers. The classifiers used here are eager learners who took more time to learn from the training data and then took less time for predicting the labeled result from the test data. Here 10-fold cross-validation is used to improve the accuracy of prediction. The output labels are classified as a pass (grade mark>=9) and fail (grade mark<9).

The steps mentioned above are typical process used in this study. The following sections provide a detailed description of each parameter and model used to construct the prediction model. The first is an evaluation analysis using regression to identify key factors that influence student growth in studies. The latter is a prediction model based on four machine learning classification algorithms.

3.4 Experimental Setup

The data set used in the proposed system is a multivariate dataset consisting of 524 instances and 30 attributes. The data is first collected and pre-processed using Microsoft Excel. Data pre-processing involves outlier detection, removing redundant values, and data integration. The data attributes include academic grades, socio-economic, and institutional physical, behavioral features of students, and it was collected by conducting surveys using questionnaires and students progress reports. The dataset was modeled using machine learning techniques.

Blockchain module

To represent blockchain and each blocks, classes of Python are used to implement member variables and methods. Methods are used to add a block, securing blocks of chain and calculate the proof of work. Dataset after pre-processing is entered into a blockchain module. The student enrollment number can be used as a unique identifier to add the student data in the block chain module. This unique ID can help track transactions between two strange entities in a digital transaction. Similarly, methods are used to calculate the hash code of the data in the blocks using Secure Hash Algorithm 256 (SHA 256). Each block contains a timestamp value and the previous block's hash code. Verification is done here by looking at the hash code contained in the current block with the hash code of the previous block. If any difference occurs in comparison, then the chain will be broken. User authentication is done through the Application Programming Interface (API) by verifying user credentials.

Predictive Analytics module

Secure data from the blockchain module is fed into the regression model. Here Linear Regression is used to find an adjective that accurately predicts a student's academic performance. The linear Regression method is widely used to find relationships between independent variables and dependent variables.

Predictive Analytics module is used to find out the variations of the G3 final year grade (dependent variable) and other educational and personal characteristics. There is a strong correlation between the qualifications of two students (G1 and G2) and the final year of high school (G3). To find out the influence of the variable dependence between each feature and the G3 range a correlation coefficient analysis is performed, it analyses collinearity between variables. To determine the influence of each variable in a student's G3 grade in the final year, we used the coefficient of determination. A heatmap is used to visualize the coefficient of correlation between attributes. Negative, positive, and neutral correlations are visualized.

After determining the appropriate attributes, the data is fed into each of the four classification models by dividing the dataset into training, validation, and test data sets. To determine the performance of each model we used the confusion matrix as a test matrix. A comparison of performance of the four models are done using certain evaluation criterias. Here 10-fold cross-validation was used to test the models, here data is divided to two parts first set contain 9 sets and second part containing the test data. This is an iterative process that will repeat ten times. Thus the number of testing observations can be increased by using the above process. Models were built using the WEKA software and using Python using the Jupyter notebook.

3.5 Evaluation Metrics

The following evaluation matrics are used to evaluate the performance of the Machine Learning Models.

The first one is accuracy it is the percentage of results that are predicted correctly. It can be measured by using the following equation:

Accuracy = (TruePositive + TrueNegative)/(TruePositive + TrueNegative + FalsePositive + FalseNegative)

The recall is the ratio of predictions that are predicted correctly as positive and are actually positive predictions that are falsely predicted as negative

Recall = True Positive/(True Positive + False Negative)

Precision is the percentage of predictions that are actually positive and predicted as positive observations

Precision = True Positive/(True Positive+ False Positive)

F1-Measure depicts the equilibrium between the Recall and the Precision metrics, and it strengthens the performance capability of a classifier.

F1Measure = 2× Recall× Precision/(Recall + Precision)

TruePositive Rate (TP) is the number of instances predicted as positive which are actually positive itself.

FalsePositive Rate (FP) is the number of instances that are actually negative but predicted as positive.

TrueNegative Rate (TN), it is the number of instances that are actually negative and also predicted as negative.

FalseNegative Rate (FN): the number of instances that are actually positive but predicted as negative.

The above parameters are determined using a confusion matrix, consisting of real values and predicted values. Table 1 shows the model performance using parameters such as accuracy, precision, recall, and

Table 1. Evaluation metrics for classification techniques in python.

	Performance Measures			
Classification Technique	**Accuracy**	**Precision**	**Recall**	**F1- measure**
ANN	93	92	93	92
SVM	91	90	92	90
RF	91	91	90	90
NB	90	89	91	89

f1 rating. These measures are applied to each prediction model and by obtaining the results model performance is evaluated in Python and WEKA.

4. Results and Discussions

This section includes two key sections to understand the proposed system. The first is about the Blockchain part of the system. The second clause describes part of the predictive statistics related to the future academic performance of the learner. This section depicts the accuracy of the corresponding models. The data set contains socio-economic, ethical, and educational aspects of students. The target class in the proposed system is the class of students, whether they pass or fail. This is a targeted feature based on the remaining elements.

4.1 Blockchain Module

To build a block, python classes are used to initialize member variables and methods. The methods are used to add a block, validate the block chain and the proof of work is then computed. Likewise blocks methods are used to compute hash code using SHA 256. Here student data records are passed into the chain in the form of a Python dictionary and the chain itself is a Python list. Genesis block is the first block in the chain which has no data in it and contains zero hash value. Each of the block includes a timestamp value and a hash code of the former block. The "proof of work" algorithm is used here to calculate hash values. Authentication is done hereby checking the hash code contained in the current block with the previous block's hashcode. If a mismatch occurs the chain will be broken. The implementation of the decentralized system enhanced the ability of a single node system to a peer to peer system. By this function we can add a new participant within the network. The flask server appends the address of the new node to be added to the known nodes list, when a post request is received from the node to be added. The node list is shared with peers of the network.

4.2 Predictive Analytics Module

Most Significant Attributes

By analyzing the correlation coefficient, important attributes are found that influence the academic performance of students. As in Table 2. WEKA provided a correlation coefficient of 0.89 between the final year's grade and the previous year's grade points. Python provided a correlation coefficient of 0.90. The results of this system show that the final year grade point is strongly influenced by previous year grade points, parent education, study time, alcohol consumption, outings, and past failure. Institutions of higher learning must consider more student features which affect future student performance like education of parents, nursery education, social network interaction, out-of-school activities, other skills, etc. By using Machine Learning techniques, educationalists can identify students at the edge of failure, this help academicians to give keen attention to them and can develop a strategy to improve their academic performance.

Many factors can affect the academic performance and growth of student knowledge. Student features are divided into different categories of social, educational, and behavioral features. Features considered in this study are student school, student gender, student age, student home address (city or rural), family size, parental status, mother-father education, mother's or father's work, school infrastructure, failure - the number of failed classes in the past, additional school support, exit, alcohol consumption, family education background, after-school activities, home internet access, absenteeism rate, etc. Last year's grades played a pivotal role in dealing with learning and knowledge growth. Locality and demographics of students also impact the learning outcomes. For example, the rank of school, classroom environment, teacher support influence their study. Another factor is related to the student profile which includes the student's personal informations. Student characteristics (psychophysical and behavioral) have a significant impact on academic performance and can predict their achievements and their necessities. In addition, assessment of students' learning and knowledge is also important in dealing with student performance assumptions. Student satisfaction in studies and course subject details can be used effectively to predict student performance. The dedication and attitude of the student are also the factors that impacted their academic performance.

The other results are:

- Past failures harm future grades.
- Alcohol abuse and students' free time are closely linked and adversely affect grade points.

Table 2. Comparison of classifier performance.

Model	WEKA (Accuracy)	Python (Accuracy)
ANN	94%	93%
SVM	91%	91%
RF	92%	91%
NB	90%	90%

- Study time is positively correlated with test results.
- Parents' education plays a significant role in student's academic performance.

Classification Technique Performance

The effectiveness of the Prediction models is tested using a confusion matrix by using the parameters such as accuracy, precision, recall, and f1 values. In WEKA, the models gave the highest accuracy with ANN having 94% accuracy as illustrated in Table 2. Figure 3 provides the classification technique performance in Python, ANN has the highest accuracy, precision, recall, and f1 measures among the four classifiers. Figure 4 provides a summary of the accuracy, TruePositive and Falsepositive values of the four classifiers in the WEKA software. The Naïve Bayes has a very high False Positive value while ANN has a very high true positive value. This figure itself indicates the most effective prediction model in this system. After using the four mentioned classifiers, we can predict the future academic performance of students and can identify students with higher risk of failures and drop-out tendencies. Evaluation measures including Accuracy, precision etc, were high for each method of classification. So,the educationalists can use the proposed method to predict student performance at an early stage of study. The ANN measurement method exceeds other classification strategies in both tested performance metrics: accuracy level (93%), representing class efficiency, and accuracy level (92%), which represents the severity of class divisions. This result is similar to the results of other studies [4]. The ANN measurement method surpasses others in recall (93%), which represents the sensitivity of the classifier, and the F1-Measure (92%), which provides the balance between recall and accuracy. The Naïve Bayes performed with low accuracy among all evaluation metrices. From the results, it is clear that the ANN classification is more efficient than the other machine learning strategies used.

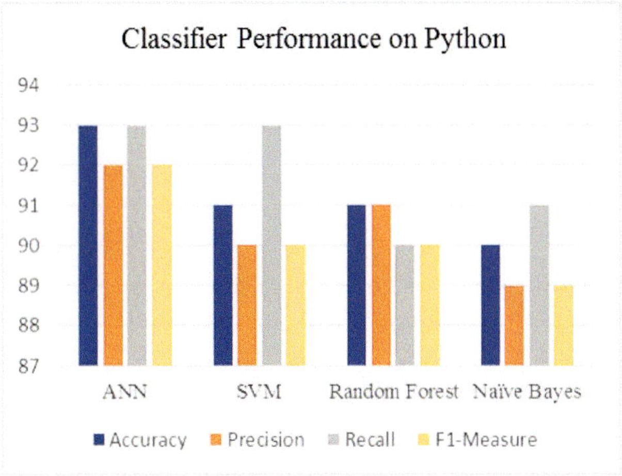

Fig. 3. Classifier performance on Python.

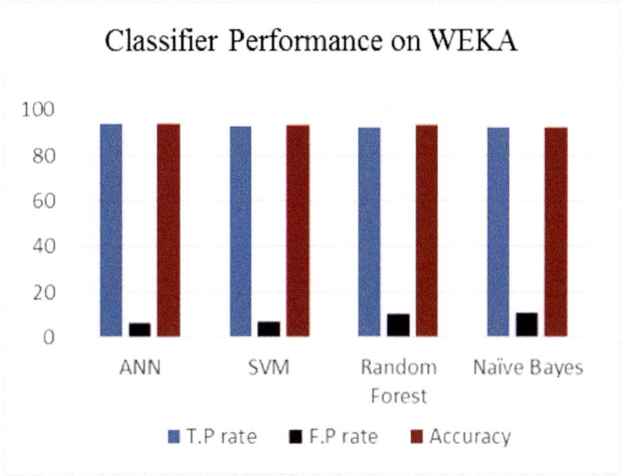

Fig. 4. Classifier performance on WEKA.

Student Categorization based on their Academic performance Prediction

Based on the output of the four prediction models we can divide students into two groups, those who are likely to succeed and those who are not. This distinction can be used effectively to reduce the failure rate and the tendency of students to drop out of higher education. Scholars can give great value to students who are weak in their studies. In this way, academic institutions can improve their fame and academic results. They can counsel students who need help to overcome their learning difficulties. Subject choices can also be made based on educational performance predictions. If a student's level of failure is at a high level in a particular subject, he can know it at the beginning of his study itself, so he may choose another subject that best suits his skills. This is how higher education institutions can mentor their students from their first year of the study itself.

5. Conclusions

The proposed system aims to support the higher education sector to develop a student future performance prediction system to identify students who are at higher risks of failures and to provide them necessary support. In the proposed system blockchain technology has been very effective in reducing the risk of data fraud for student data records. It explains how educational institutions can capture student data blocks and provide much-needed assistance based on future forecasts. The predictive analytics module predicts future academic performance for students to improve their academic performance. Four speculative models are proposed and developed for prediction, they are, Artificial Neural Network, Random Forest, Support Vector Machine, and Naïve Bayes classifiers. The attribute set contains student grades, behavioral, school factors, physical, mental features, and were compiled using school reports and questionnaires. The methodology adapted in the proposed system is generalizable, so it can be applied in any academic institutions. This particular scheme confirms the efficiency of predictive Analytics models in educational sector where experts can use the models to plan and manage the availability of limited institutional resources for educational enhancement. In addition, the results show the effectiveness of early student performance prediction system based on academic and student social and behavioral characteristics. The secure storage of student data provided the data security required by each educational institutions for educational data analysis. Here, ANN model achieved a performance accuracy of 93%. Furthermore, the results of this study show that the ANN method is superior in both accuracy and precision. The Naïve Bayes model gave the worst results. Using the proposed model educational institutions can find out students who have chances to fail and drop out, which will help give them more attention to improve their academic performance. Some of the studies have explicitly used WEKA to predict, therefore, so a comparative analysis of the results obtained in both WEKA and Python are performed. The results showed that the WEKA software is reliable. The WEKA software provided more accurate predictions than our Python-trained model. Accuracy, accuracy, recall, and F1 rating for each model were high in WEKA.

References

[1] Abu Zohair, L.M. 2019. Prediction of Student's performance by modelling small dataset size. Int. J. Educ. Technol. High. Educ., 16(1). doi: 10.1186/s41239-019-0160-3.

[2] Adekitan, A.I. and Salau, O. 2020. Toward an improved learning process: The relevance of ethnicity to data mining prediction of students' performance. SN Appl. Sci., 2(1): 1–15. doi: 10.1007/s42452-019-1752-1.

[3] Awaji, B., Solaiman, E. and Albshri, A. 2020. Blockchain-based applications in higher education: A systematic mapping study. July, 2020. doi: 10.1145/3411681.3411688.

[4] Sharples, M. and Domingue, J. 2016. The Blockchain and Kudos: A distributed system for educational record, reputation and reward, 2: 490–496. doi: 10.1007/978-3-319-45153-4.

[5] Reis-Marques, C., Figueiredo, R. and de C. Neto, M. 2021. Applications of blockchain technology to higher education arena: A bibliometric analysis. Eur. J. Investig. Heal. Psychol. Educ., 11(4): 1406–1421. doi: 10.3390/ejihpe11040101.

[6] Alzahrani, B., Bahaitham, H., Andejany, M. and Elshennawy, A. 2021. How ready is higher education for Quality 4.0 transformation according to the LNS Research Framework, pp. 1–29.

[7] Mengash, H.A. 2020. Using data mining techniques to predict student performance to support decision making in university admission systems. IEEE Access, 8: 55462–55470. doi: 10.1109/ACCESS.2020.2981905.

[8] Sothan, S. 2019. The determinants of academic performance: Evidence from a Cambodian University. Stud. High. Educ., 44(11): 2096–2111. doi: 10.1080/03075079.2018.1496408.

[9] Yang, S.J.H., Lu, O.H.T., Huang, A.Y.Q. Huang, J.C.H., Ogata, H. and Lin, A.J.Q. 2018. Predicting students' academic performance using multiple linear regression and principal component analysis. J. Inf. Process., 26: 170–176. doi: 10.2197/ipsjjip.26.170.

[10] Polyzou, A. and Karypis, G. 2016. Grade prediction with models specific to students and courses. Int. J. Data Sci. Anal., 2(3-4): 159–171. doi: 10.1007/s41060-016-0024-z.

[11] Hu, Q. and Rangwala, H. 2019. Academic performance estimation with attention-based graph convolutional networks. EDM 2019 - Proc. 12th Int. Conf. Educ. Data Min., pp. 69–78.

[12] Yao, Y., Zhang, Z., Cui, H., Ren, T. and Xiao, J. 2019. The influence of student abilities and high school on student growth: A case study of chinese national college entrance exam. IEEE Access, 7: 148254–148264. doi: 10.1109/ACCESS.2019.2946503.

[13] Miguéis, V.L., Freitas, A., Garcia, P.J.V. and Silva, A. 2018. Early segmentation of students according to their academic performance: A predictive modelling approach. Decis. Support Syst., 115: 36–51. doi: 10.1016/j.dss.2018.09.001.

[14] Nieto, Y., Gacia-Diaz, V., Montenegro, C., Gonzalez, C.C. and Gonzalez Crespo, R. 2019. Usage of machine learning for strategic decision making at higher educational institutions. IEEE Access, 7: 75007–75017. doi: 10.1109/ACCESS.2019.2919343.

[15] Amrieh, E.A., Hamtini, T. and Aljarah, I. 2016. Mining educational data to predict student's academic performance using ensemble methods. Int. J. Database Theory Appl., 9(8): 119–136. doi: 10.14257/ijdta.2016.9.8.13.

Chapter 12

Conclusion and Future Enhancement

Conclusion

Blockchain technology is one of the advanced technologies in the combined field of data structure and cryptography. It was introduced in the year of 2008 by Satoshi Nakamoto through cryptocurrency. At present this technology has been used in almost all the fields like finance, business, supply chain, healthcare, etc. It contains list of records in the form of blocks with transparency decentralized characteristics. Many of the applications related to blockchain focus on interoperability between privacy and security. Many healthcare applications contain the property of Immutability in blockchain ledger for keeping unaltered and unchanged data in the system.

This book discussed about various health care applications along with threats in health care systems. Many chapters discuss about the health care systems in the aspect of blockchain technology. The health care applications are directly and indirectly supported and benefitted by block chain. The benefits are as given below

1. Provide security to patient's personal and medical data
2. Provide security to the clinical trials
3. Provide Quality management in the service and facilities
4. Supply chain management for medicine
5. Better way to support diagnosis to treatment.

In this book many chapters focus on health care system for improving the confidentiality about patient data like personal and health data, summarized the threats and applications in blockchain. The chapters discuss about the difficulties in implementing this technology and the social aspect of blockchain in the society.

Future Enhancement

One of the major challenges in the healthcare sector is to improve the patient care. In the present scenario, healthcare system is facing a lot of challenges in terms of cost and quality. Most of the diagnoses and treatment methods always follow the same protocol for every patient.

In the future there should be more research focussing on the issues like affordable cost, secure and precise diagnosis, provide right treatment at right time. In the blockchain technology smart health care is referred as implementing good health care support along with computational tools and IoT interface.

Digitalisation

Digitalisation is an important aspect of healthcare in the future. Machine learning, IoT, and Artificial Intelligence will be integrated to provide solutions for the existing issues. All the heath care data are efficiently processed for prediction and diagnoses of major diseases in the initial stage. The computational tools support major changes in the treatment procedures. The healthcare industry is forced to do this for more cost benefits. The main scope for future enhancement is to improve performance of the tool in a cost-effective manner.

It should provide Block chain concept cooperate with Recent advance IoT technology:

Improvement in the access of health care devices for both patients and doctors.

Provide 24/7 remote monitoring usinf IoT enabled devices.

Increasing accuracy in the measurement of digitalized devices.

Provide Artificial intelligence and machine learning tools for better understanding and analysing the illness of the patients.

This improvement in the patient experience provides good health and security.

Index

M

P

R

S

T